INTRODUCING
ISSUES WITH
OPPOSING
VIEWPOINTS®

Energy Alternatives

Lauri S. Friedman, *Book Editor*

GREENHAVEN PRESS
A part of Gale, Cengage Learning

GALE
CENGAGE Learning™

Detroit • New York • San Francisco • New Haven, Conn • Waterville, Maine • London

Christine Nasso, *Publisher*
Elizabeth Des Chenes, *Managing Editor*

© 2011 Greenhaven Press, a part of Gale, Cengage Learning

For more information, contact:
Greenhaven Press
27500 Drake Rd.
Farmington Hills, MI 48331-3535
Or you can visit our Internet site at gale.cengage.com

For product information and technology assistance, contact us at

Gale Customer Support, 1-800-877-4253
For permission to use material from this text or product, submit all requests online at
www.cengage.com/permissions

Further permissions questions can be e-mailed to permissionrequest@cengage.com

Articles in Greenhaven Press anthologies are often edited for length to meet page requirements. In addition, original titles of these works are changed to clearly present the main thesis and to explicitly indicate the author's opinion. Every effort is made to ensure that Greenhaven Press accurately reflects the original intent of the authors. Every effort has been made to trace the owners of copyrighted material.

Cover image copyright © by J.D.S./Shutterstock.com.

LIBRARY OF CONGRESS CATALOGING-IN-PUBLICATION DATA

Energy alternatives / Lauri S. Friedman, book editor.
      p. cm. -- (Introducing issues with opposing viewpoints)
   Includes bibliographical references and index.
   ISBN 978-0-7377-5198-7 (hardcover)
   1. Power resources. 2. Renewable energy sources. I. Friedman, Lauri S.
   TJ163.24.E526 2011
   333.79--dc22

                                                                  2011000872

Printed in the United States of America
1 2 3 4 5 6 7 15 14 13 12 11

Contents

Foreword

Indulging in a wide spectrum of ideas, beliefs, and perspectives is a critical cornerstone of democracy. After all, it is often debates over differences of opinion, such as whether to legalize abortion, how to treat prisoners, or when to enact the death penalty, that shape our society and drive it forward. Such diversity of thought is frequently regarded as the hallmark of a healthy and civilized culture. As the Reverend Clifford Schutjer of the First Congregational Church in Mansfield, Ohio, declared in a 2001 sermon, "Surrounding oneself with only like-minded people, restricting what we listen to or read only to what we find agreeable is irresponsible. Refusing to entertain doubts once we make up our minds is a subtle but deadly form of arrogance." With this advice in mind, Introducing Issues with Opposing Viewpoints books aim to open readers' minds to the critically divergent views that comprise our world's most important debates.

Introducing Issues with Opposing Viewpoints simplifies for students the enormous and often overwhelming mass of material now available via print and electronic media. Collected in every volume is an array of opinions that captures the essence of a particular controversy or topic. Introducing Issues with Opposing Viewpoints books embody the spirit of nineteenth-century journalist Charles A. Dana's axiom: "Fight for your opinions, but do not believe that they contain the whole truth, or the only truth." Absorbing such contrasting opinions teaches students to analyze the strength of an argument and compare it to its opposition. From this process readers can inform and strengthen their own opinions, or be exposed to new information that will change their minds. Introducing Issues with Opposing Viewpoints is a mosaic of different voices. The authors are statesmen, pundits, academics, journalists, corporations, and ordinary people who have felt compelled to share their experiences and ideas in a public forum. Their words have been collected from newspapers, journals, books, speeches, interviews, and the Internet, the fastest growing body of opinionated material in the world.

Introducing Issues with Opposing Viewpoints shares many of the well-known features of its critically acclaimed parent series, Opposing Viewpoints. The articles are presented in a pro/con format, allowing readers to absorb divergent perspectives side by side. Active reading questions preface each viewpoint, requiring the student to approach the material thoughtfully and carefully. Useful charts, graphs, and cartoons supplement each article. A thorough introduction provides readers with crucial background on an issue. An annotated bibliography points the reader toward articles, books, and websites that contain additional information on the topic. An appendix of organizations to contact contains a wide variety of charities, nonprofit organizations, political groups, and private enterprises that each hold a position on the issue at hand. Finally, a comprehensive index allows readers to locate content quickly and efficiently.

Introducing Issues with Opposing Viewpoints is also significantly different from Opposing Viewpoints. As the series title implies, its presentation will help introduce students to the concept of opposing viewpoints and learn to use this material to aid in critical writing and debate. The series' four-color, accessible format makes the books attractive and inviting to readers of all levels. In addition, each viewpoint has been carefully edited to maximize a reader's understanding of the content. Short but thorough viewpoints capture the essence of an argument. A substantial, thought-provoking essay question placed at the end of each viewpoint asks the student to further investigate the issues raised in the viewpoint, compare and contrast two authors' arguments, or consider how one might go about forming an opinion on the topic at hand. Each viewpoint contains sidebars that include at-a-glance information and handy statistics. A Facts About section located in the back of the book further supplies students with relevant facts and figures.

Following in the tradition of the Opposing Viewpoints series, Greenhaven Press continues to provide readers with invaluable exposure to the controversial issues that shape our world. As John Stuart Mill once wrote: "The only way in which a human being can make some approach to knowing the whole of a subject is by hearing what can be said about it by persons of every variety of opinion and studying all modes in which it can be looked at by every character of mind. No wise man ever acquired his wisdom in any mode but this." It is to this principle that Introducing Issues with Opposing Viewpoints books are dedicated.

Introduction

Environmental, political, and geological problems associated with fossil fuel use have stimulated interest in—and debate over—alternative energy sources for decades. But in recent years, economic hardship caused by the global recession has generated debate over whether transitioning to a clean energy economy could create much-needed jobs while solving long-standing problems with fossil fuel use. Some argue that transitioning to energy alternatives will create a new class of "green-collar" jobs that are good for both the environment and the economy; others contend that not only will green-collar jobs not significantly boost the economy, they will actually cause overall job loss.

According to Van Jones, author of *The Green Collar Economy: How One Solution Can Fix Our Two Biggest Problems*, a green-collar job is "a blue-collar job that is graded to better respect the environment. It's a job where your wealth is not taking away from the community's or the planet's health."[1] Green jobs have been developed in nearly every market sector. They include green interior designers (who design homes or businesses with energy-efficient products, furniture, flooring, and windows), green landscape architects (who landscape with native and water-appropriate plants), and green fashion designers (who work with organic cotton, bamboo, and other sustainable fibers). Other green jobs are related to the production of hybrid or electric vehicles, solar and wind power development, and other alternative energy technologies that require cutting-edge skills and materials.

Growth in these clean energy sectors is hoped to spur job creation and simultaneously help the environment and the economy. For example, the *Boston Globe* reports that as energy-saving air-conditioners and refrigerators become more desirable and affordable, the need for specially trained mechanics and installation personnel is expected to increase dramatically. So too is the need for environmental engineers: The Bureau of Labor Statistics (BLS) projects that by 2014, the impending green economy will open 31.2 percent more jobs for environmental engineers than were needed in 2004. Protection technicians—people who monitor pollution levels in air, water, and land—are also expected to be in demand: The BLS estimates the need for protection technicians will increase by 28 percent. Organic farmers and agricultural managers, green investment analysts, and fuel cell engineers are just a few of the other jobs that the government expects to be created as the economy goes increasingly green.

That a greener economy could drive job creation was one of the bases of President Barack Obama's 2009 stimulus package, known as the American Recovery and Reinvestment Act of 2009 (ARRA). ARRA earmarked more than $92 billion in spending and tax breaks for clean energy–related programs. According to the Progressive States Network, "Since the ARRA was enacted, an estimated 150,000 jobs were saved or created in the construction of solar panels and wind turbines."[2]

Adopting standards that would require states to generate a certain portion of their energy from renewable resources could add further jobs to the economy. According to the Union of Concerned Scientists, adopting standards that would require states to get at least 20 percent of their energy from renewable resources by 2020 could create 355,000 American jobs. A different study, by Navigant Consulting, found that if standards were increased to 25 percent by 2025, the number of clean energy jobs available in the wind, solar, hydropower, biomass, and other renewable energy industries would more than double.

But many remain doubtful of the green economy's ability to add a meaningful number of jobs to the national economy. For one, more than 8 million Americans have lost their jobs since the recession began in 2008. Thus millions of jobs need to be added back into the market to restore pre-2008 economic conditions. "The green sector is simply not large enough or competitive enough to be a major engine of job creation,"[3] says Samuel Sherraden, a policy analyst with the New America Foundation. Sherraden and others argue that green energy sources like wind and solar power are not capable of meeting the bulk of America's energy needs and are therefore incapable of generating anything more than a small number of jobs in those industries. Evidence from other countries indicates that the creation of green jobs can possibly result in a net loss of jobs. A widely cited

2009 report by professor Gabriel Calzada revealed that in Spain—the largest national supporter of renewable energy electricity—jobs that were created by renewable energy industry growth were temporary and heavily subsidized. Columnist George F. Will reports that alternative energy jobs "have received $752,000 to $800,000 each in subsidies—wind industry jobs cost even more, $1.4 million each."[4] Worse, Calzada calculated that each renewable energy job resulted in the loss or lack of creation of 2.2 other jobs in other industries. The study concluded that the creation of clean energy jobs actually resulted in a total loss of about 110,000 Spanish jobs. "Killing jobs in efficient industries to create jobs in inefficient ones is hardly a recipe for economic success," says Max Schulz, an analyst at the Manhattan Institute. "There may be legitimate arguments for taking dramatic steps to fight climate change. Boosting the economy isn't one of them."[5]

On the other hand, it has been argued that many green jobs have in fact been created by the 2009 stimulus—just not *American* jobs. Indeed, the US Department of Energy has admitted that more than $2 billion in tax credits were given to companies that use Chinese, South Korean, and Spanish workers for manufacturing jobs, rather than American workers. Similarly, a 2009 report by American University's Investigative Reporting Workshop found that eleven American wind farms that received stimulus money used it to purchase hundreds of wind turbines from foreign manufacturers, rather than American ones. Said Sherraden, "It is impossible to guarantee that clean-energy stimulus is not leaked abroad. We have to recognize that we are funding job-creation programs in Germany, Spain, Japan and China."[6]

Whether clean energy jobs can shuttle the United States into healthier economic times remains to be seen, and that is just one of the issues debated in *Introducing Issues with Opposing Viewpoints: Energy Alternatives*. Readers will also consider whether alternatives to fossil fuels are necessary, efficient, viable, and environmentally friendly. Guided reading questions and thought-provoking essay prompts help engage students with the material and encourage them to form their own opinions on this increasingly relevant matter.

Notes

1. Quoted in Maria José Viñas, "An Interview with Van Jones," Mong abay.com, June 23, 2008. http://news.mongabay.com/2008/0623-van_jones_interview_mj_ucsc.html.
2. Fabiola Carrion, "Clean Energy Options: In the Wake of the Oil Spill, Energy Alternatives That Will Create Jobs," Progressive States Network, July 19, 2010. http://progressivestates.org/node/25318.
3. Samuel Sherraden, "Green Jobs: Hope or Hype?," CNN.com, July 28, 2009. http://articles.cnn.com/2009-07-28/politics/sher raden.green.jobs_1_green-jobsgreen-sector-energy-efficiency?_s=PM:POLITICS.
4. George F. Will, "Tilting at Green Windmills," *Washington Post*, June 25, 2009. www.washingtonpost.com/wp-dyn/content/arti cle/2009/06/24/AR2009062403012.html.
5. Max Schulz, "Don't Count on 'Countless' Green Jobs: The Evidence Shows Alternative Energy Is Expensive," *Wall Street Journal*, February 20, 2009. http://online.wsj.com/article/SB123509599682529113.html.
6. Quoted in Patrice Hill, "'Green' Jobs No Longer Golden in Stimulus," *Washington Times*, September 8, 2010. www.washing tontimes.com/news/2010/sep/9/green-jobs-no-longer-golden-in-stimulus/?page=1.

Chapter 1

Is It Necessary to Transition to Alternative Sources of Energy?

This oil production platform in the North Sea represents the current energy source, for which alternatives are being sought.

Oil Reserves Are Running Out

Nicholas C. Arguimbau

"The supply of the world's most essential energy source is going off a cliff."

Nicholas C. Arguimbau is an environmental lawyer. In the following viewpoint he warns that oil reserves are running out faster than previously expected. He explains that production of oil has been steadily dropping and is poised to drastically decline in just a few years. Arguimbau says the situation is made worse by the fact that demand for oil is up—nations such as China and India are causing oil to be consumed at an ever faster rate, which is depleting supply faster than expected. Because there is no energy source to replace oil, Arguimbau believes that Americans—and consumers of oil everywhere—are in for a terrible shock. He advises communities to begin preparing for the end of oil, which he believes will be sudden and catastrophic.

AS YOU READ, CONSIDER THE FOLLOWING QUESTIONS:
1. In how many years will conventional oil be almost all gone, according to Arguimbau?
2. In what year does Arguimbau warn demand for oil will begin to outstrip supply?
3. Name at least five situations the author warns Americans have zero time to plan for.

Nicholas C. Arguimbau, "The Imminent Crash of Oil Supply: Be Afraid," *CounterCurrents.org,* April 23, 2010. Reprinted by permission.

Look at this graph and be afraid. It does not come from [the environmental group] Earth First. It does not come from the Sierra Club. It was not drawn by Socialists or Nazis or Osama Bin Laden or anyone from Goldman-Sachs. If you are a Republican Tea-Partier, rest assured it does not come from a progressive Democrat. And vice versa. It was drawn by the United States Department of Energy, [DOE] and the United States military's Joint Forces Command concurs with the overall picture.

Oil Is Running Out—and Fast

What does it imply? The supply of the world's most essential energy source is going off a cliff. Not in the distant future, but in a year and a half [from April 2010]. Production of all liquid fuels, including oil, will drop within 20 years to half what it is today. And the difference needs to be made up with "unidentified projects," which one of the world's leading petroleum geologists says is just a "euphemism for rank shortage," and the world's foremost oil industry banker says is "faith based."

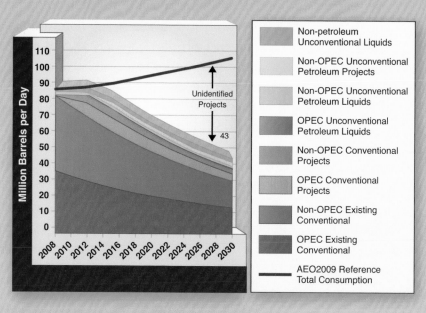

Taken from: US Energy Information Administration, "Annual Energy Outlook 2009."

This graph was prepared for a DOE meeting in spring, 2009. Take a good look at what it says, assuming it to be correct:

1. Conventional oil will be almost all gone in 20 years, and there is nothing known to replace it.
2. Production of petroleum from existing conventional sources has been dropping at a rate slightly over 4% per year for at least a year and will continue to do so for the indefinite future.
3. The graph implies that we are past the peak of production and that there are 750 billion barrels of conventional oil left (the areas under the "conventionals" portion of the graph, extrapolated to the right as an exponential). Assuming that the remaining reserves were 900 billion or more at the halfway point, then we are at least 150 billion barrels, or 5 years, past the midpoint.
4. Total petroleum production from all presently known sources, conventional and unconventional, will remain "flat" at approximately 83 mbpd [million barrels per day] for the next two years and then will proceed to drop for the foreseeable future, at first slowly but by 4% per year after 2015.

When Demand Is Greater than Supply

5. Demand will begin to outstrip supply in 2012, and will already be 10 million barrels per day [bpd] above supply in only five years. The United States Joint Forces Command concurs with these specific findings. 10 million bpd is equivalent to half the United States' entire consumption. To make up the difference, the world would have to find another Saudi Arabia and get it into full production in five years, an impossibility.
6. The production from presently existing conventional sources will plummet from its present 81 mbpd to 30 mbpd by 2030, a 63% drop in a 20-year period.
7. Meeting demand requires discovering, developing, and bringing to full production 60 mbpd (105–45) of "unidentified projects" in the 18-year period of 2012–2030 and approximately 25 mbpd of such projects by 2020, on the basis of a very conservative estimate of only 1% annual growth in demand. The independent Oxford Institute of Energy Studies has estimated a possible development of 6.5 mbpd of such projects, including the

Canadian tar sands, implying a deficit of 18–19 mbpd as compared to demand, and an approximate 14 mbpd drop in total liquid fuels production relative to 2012, a 16% drop in 8 years.

8. The curve is virtually identical to one produced by geologists Colin Campbell and Jean Laherrere and published in "The End of Cheap Oil," in *Scientific American*, March, 1998, twelve years ago. They projected that production of petroleum from conventional sources would drop from 74 mbpd in 2003 (as compared to 84 mbpd in 2008 in the DOE graph) and drop to 39 mbpd by 2030 (as compared to 39 mbpd by 2030 in the DOE graph!). Campbell and Laherrere predicted a 2003 "peak," and the above graph implies a "peak" (not necessarily the actual peak, but the midpoint of production) of 2005 or before.

An oil exploration rig looks for oil off the coast of Sumatra. The demand for oil is forcing exploration for new sources all over the world.

Americans Think Oil Is Running Out

The majority of Americans think oil reserves are running out and want access to alternative sources of energy.

Question:
Do you think that governments should make long-term plans based on the assumption that

23%	76%
enough new oil will be found so that it can remain a primary source of energy for the foreseeable future	oil is running out and it is necessary to make a major effort to replace oil as a primary source of energy

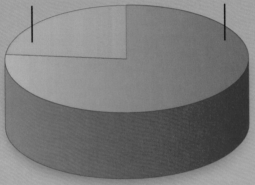

Taken from: January 2008 WorldPublicOpinion.org.poll.

Zero Time to Plan

So here we are, if the graph is right, on the edge of a precipice, with no prior warning from either the industry, which knows what it possesses, or the collective governments, which ostensibly protect the public interest. As Colin Campbell, a research geologist who has worked for many large oil companies and studied oil depletion extensively says, "The warning signals have been flying for a long time. They have been plain to see, but the world turned a blind eye, and failed to read the message." The world was completely transformed by oil for the duration of the twentieth century, but if the graph is right, within 20 years it will be virtually gone but our dependence upon it will not. Instead, we have

- zero time to plan how to replace cars in our lives.
- zero time to plan how to manufacture and install milions of furnaces to replace home oil furnaces, and zero time to produce the infrastructure necessary to carry out that task.
- zero time to retool suburbia so it can function without gasoline.
- zero time to plan for replacement of the largest military establishment in history, almost completely dependent upon oil.
- zero time to plan to support nine billion people without the "green revolution," a creation of the age of oil.
- zero time to plan to replace oil as an essential fuel in electricity production.
- zero time to plan for preserving millions of miles of roads without asphalt.
- zero time to plan for the replacement of oil in its essential role in EVERY industry.
- zero time to plan for replacement of oil in its exclusive role of transporting people, agricultural produce, manufactured goods. In a world without oil that appears only twenty years away, there will be no oil-burning ships transporting US grain to other countries, there will be no oil-burning airlines linking the world's major cities, there will be no oil-burning ships transporting Chinese manufactured goods to the billions now dependent on them.
- zero time to plan for the survival of the billions of new people expected by 2050 in the aftermath of "peak everything."
- zero capital, because of failing banks and public and private debt, to address these issues.

> # FAST FACT
>
> In 1956 US geophysicist M. King Hubbert predicted that oil production in the continental United States would peak between 1965 and 1970. Oil production peaked in that region in 1970–1971.

Why No Time to Plan?

Because if we at any time use more oil than allowed by the graph, we will have even less later.

Because we are already committed to supporting 2.5 billion more people on what we have.

Because every day we continue upward in our oil consumption, even though we continue to have more people who need it and billions who deserve to rise from abject poverty, we are making the future supply shortage worse.

If you believe the graph, demand will outstrip supply starting at the end of 2011, and severely outstrip supply in five years. What are we going to do, and how are we going to do it? We have no time to decide.

A Fast Ride Down

It is very unlikely that things can be better than the graph indicates. Why?

The great majority of authorities believe there is little more than 1 trillion barrels of conventional oil left. You can make a simple calculation from that: At the present rate of 30 billion barrels per year, 82 million barrels per day, it will all be gone in 33 years, and consumption has been rapidly increasing, not decreasing, so if anything it will all be gone sooner. . . .

A closer look at the graph reveals that it was drawn on the assumption that the world's existing conventional fields contain only 750,000 barrels at this time, enough to keep us going only 25 years.

The graph assumes a decline rate of 4% per year. As long as the estimates of remaining reserves are right, that can't be far off. In fact, 4% is a relatively low decline rate compared to what has been observed in oil fields generally. Hold on, it's going to be a fast ride down! . . .

The End Is Near

We are on our own. We are rapidly going to have to deal with less and less oil, since there has been no forewarning and no planning. It is a time for communities to prepare for community energy independence, because only that way will we be safe. This means relying on the sun and wind and water that have always been with us. It means cooperation with each other to get through seriously difficult times. It means finding alternatives to oil throughout our lives as quickly as possible—the oil that runs our cars, the oil that heats our houses, the oil that runs generators for our electricity, the oil from which chemical fertilizers and insecticides and plastics and polyester are made, the oil that brings countless manufactured goods to us from overseas, the oil on which farmers depend for irrigation pumps, for transporting produce to market, for working the soil to bring us food. If you believe the graph, it will almost all be gone in 20 years.

EVALUATING THE AUTHORS' ARGUMENTS:

Newt Gingrich and Steve Everley, authors of the next viewpoint, call people such as Arguimbau "environmental extremists" and imply they are unnecessarily alarmist. After reading Arguimbau's viewpoint, how would you characterize his writing? Would you describe it as measured and fact-based, or would you describe it as alarmist and extreme? Explain your reasoning.

Oil Reserves Are Not Running Out

Newt Gingrich and Steve Everley

"There have been more than 200 new oil discoveries around the world this year alone."

In the following viewpoint, Newt Gingrich and Steve Everley argue that oil reserves are not running out. They take issue with the theory of peak oil, which states that humans will eventually run out of oil, a finite resource. Gingrich and Everley explain that various parties have warned about peak oil for decades. But in that same amount of time, the known number of oil reserves has only grown. They argue the earth has plenty of oil and America should craft an energy policy that is not based on the fear of running out of oil. Gingrich was Speaker of the House of Representatives from 1995 to 1999. He is currently the general chairman of American Solutions, where Everley is energy policy manager.

AS YOU READ, CONSIDER THE FOLLOWING QUESTIONS:

1. How many billions of barrels of oil do the authors say might be off the coast of Brazil? What piece of their argument does this support?
2. How many billions of barrels of oil might be in the Chukchi Sea, according to Gingrich and Everley?
3. What, according to the authors, caused the US Geological Survey to revise its estimate of oil reserves in North Dakota and Montana?

Newt Gingrich and Steve Everley, "Peak Oil: A Theory Running Out of Gas," *Investor's Business Daily,* October 8, 2009. Reprinted by permission.

One year ago [in 2008], Congress responded to the chorus of Americans calling for more American energy by lifting the ban on offshore drilling. For the first time in a quarter-century, it became legal to drill for more oil and natural gas reserves offshore. This anniversary allows us to look back on how far we have come since 2008. The sad reality is we have barely moved.

The Myth of Peak Oil

Earlier this year [2009], Secretary of the Interior Ken Salazar announced he would delay the comment period for offshore energy exploration by six months. Salazar claimed that the previous comment period, which would have ended in March, "by no means provides enough time for public review."

Evidently 25 years of delays and bans was not enough. During that quarter-century Congress had to make the decision each year whether to renew the ban on offshore energy, yet Salazar suggested that we were somehow engaged in a "headlong rush" to explore for energy offshore.

One reason behind this bureaucratic delay has nothing to do with developing a responsible energy policy. It has to do with the myth known as "peak oil."

Peak oil was a theory developed decades ago that suggests we will soon reach a point of maximum oil production, after which oil will only become harder and harder to find, leading to an enormous energy crisis.

An oil platform lies off the coast of Rio de Janeiro. Brazil's offshore oil deposits may reach 100 billion barrels.

In fact, many still believe this theory today, including [former vice president] Al Gore, who told CNN that "we are almost certainly at or near what they call peak oil." The Sierra Club's executive director, Carl Pope, once warned that peak oil could come in 2010 and that "we're better off without cheap gas."

Since anti-energy elites ignore the massive amounts of oil that we do have but are banned from extracting, they propose new energy taxes to supposedly save us from future energy crises by punishing the use of oil. After all, if oil is the problem, then coercing America away from oil usage would be the answer.

The problem is that peak oil is fundamentally wrong.

Constantly Discovering More Oil

Geophysicist Marion King Hubbert first suggested in 1956 that peak oil was a reality, and that we would hit our maximum rate of production sometime around 1970. But recent estimates of oil are actually an astounding three times larger than peak oil predictions, meaning the newest discoveries simply should not exist according to the theory of peak oil.

In Brazil, there could be as much as 100 billion barrels of oil offshore, including the Tupi oil field, which is the largest oil discovery in this hemisphere in 30 years. Had Brazilians been banned from exploring and conducting new seismic tests, they never would have made this massive discovery. Now Brazil is set to become an oil exporter.

Researchers from the U.S. Geological Survey [USGS] concluded earlier this year that there are massive amounts of oil and natural gas in the Chukchi Sea off Alaska's coast. They estimated that there could be as much as 157 billion barrels of oil in the Arctic, or nearly twice as much oil as was previously known to exist in that part of the world. The natural gas discovery is also greater than all of the previously known reserves in the Arctic.

Last year the USGS had to increase its estimate of oil reserves in the Bakken formation in North Dakota and Montana by 2,500%. The area is now estimated to hold more than 4 billion barrels of oil.

Hundreds of New Discoveries Are Made Each Year

In Israel, experts underestimated the size of a huge natural gas discovery made in January of this year. The field is actually 16% larger than what had been estimated. Experts now claim Israel can supply itself with

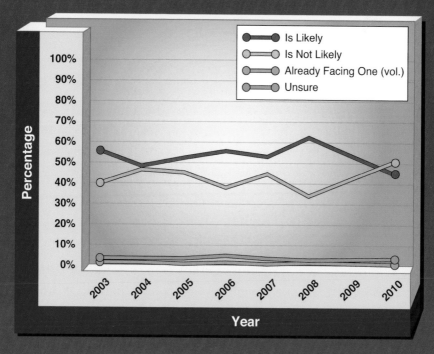

The United States Not Likely to Face an Energy Shortage

Fewer and fewer Americans believe a critical energy shortage is on the horizon.

Question:
Do you think that the United States is or is not likely to face a critical energy shortage during the next five years?

Taken from: Gallup poll, March 4–7, 2010.

enough natural gas for two decades and could be an energy exporter.

In the United States, we have a 100-year supply of natural gas. Last year geologists discovered that gas reserves in the Marcellus Shale formation in Appalachia are actually 250 times larger than they estimated in 2002.

And recently, in the Gulf of Mexico, BP announced they had made a huge new discovery of oil, estimated to be as large as the biggest oil-producing spots in the Gulf, which means it could supply as much as 300,000 barrels of oil per day.

No Need to Transition to Alternatives

All told, there have been more than 200 new oil discoveries around the world this year alone. What these discoveries mean is our energy future does not have to be dictated by OPEC [Organization of the Petroleum Exporting Countries] or energy taxes on American businesses. It is possible to have abundant and reliable sources of low-cost energy.

This runs contrary to what environmental extremists claim, namely that we have to make a painful transition to alternative fuels and renewables to avoid the disastrous effects of peak oil. In reality, we have reached the end of peak oil as a theory.

EVALUATING THE AUTHORS' ARGUMENTS:

Gingrich and Everley argue that although oil is a finite resource, humans are not likely to run out of it any time soon. They therefore believe the notion of peak oil is a myth, a scare tactic used by environmentalists. Nicholas C. Arguimbau, author of the previous viewpoint, however, disagrees. He believes peak oil is an impending threat, a disastrous state we are on the brink of achieving. After reading both viewpoints, state whether you think the theory of peak oil is real or imagined. Which pieces of evidence helped you make your decision?

Viewpoint

3

Energy Alternatives Are Needed to Avert Climate Change

Sarah van Gelder, Madeline Ostrander, and Doug Pibel

"People are energized by the prospect of a green economy and . . . want the opportunity not only to avert a climate catastrophe, but to help build a better future."

In the following viewpoint Sarah van Gelder, Madeline Ostrander, and Doug Pibel warn that the world's continued reliance on fossil fuels is causing climate change. The authors say that carbon emissions—climate-changing gases emitted when fossil fuels such as oil and coal are burned—have risen with each decade. These emissions are responsible for catastrophic floods, droughts, dust storms, fires, and other severe weather events that kill, wound, and displace millions of people and cost billions of dollars. The authors predict these catastrophes will worsen until humanity switches to renewable sources of energy like solar, wind, hydro, and biomass power. These energy sources burn cleanly and do not emit climate-changing gases. Van Gelder, Ostrander, and Pibel conclude

Sarah van Gelder, Madeline Ostrander, and Doug Pibel, "Climate Action: What Will It Take to Avert Disastrous Climate Change?," *Yes! Magazine*, vol. 52 (*Be a Climate Hero*), Winter 2010. Reprinted with permission.

that the only way to save the world from catastrophic climate change is to replace fossil fuels with renewable energy alternatives—and the sooner the better.

Van Gelder is cofounder and executive editor of *Yes! Magazine,* where this viewpoint was originally published. *Yes!* also employs Ostrander as a senior editor and Pibel as a contributing writer.

AS YOU READ, CONSIDER THE FOLLOWING QUESTIONS:
1. By what percent have carbon emissions grown per year since 2000, according to the authors?
2. What do the authors warn might be the fate of Lakes Mead and Powell?
3. According to a poll by Public Agenda, what percent of Americans think that investing in alternative energy sources is the best way to stimulate the economy?

For nearly any major disaster—natural, economic, or military—there was a moment when tragedy could have been prevented. In just the last decade, experts warned that a subprime mortgage bubble could lead to financial collapse and that a hurricane could devastate New Orleans. But our leaders failed to head off disaster, and the public knew little until it was too late.

Now we face the largest potential Katrina the world has ever seen, an imminent catastrophe we refer to blandly as "climate change." Neither your mayor, nor your senator, nor certainly, your president has declared a climate emergency. But in the time since you may have watched *An Inconvenient Truth,*[1] global emissions have worsened, and the scientific predictions have become much more frightening.

Climate Change Is Already Happening
The carbon dioxide that we have already put into the atmosphere makes it a near certainty that our oceans will become steadily more acidic, eventually destroying coral reefs and sea life. Glaciers will continue to melt year by year, eventually threatening the water supply of

1. A 2006 documentary on climate change.

as much as 25 percent of the human population. Sea levels are already rising, and will continue to rise for hundreds of years.

In many parts of the world, the climate emergency has already arrived. An estimated 26 million people have already been displaced by the increases in hurricanes, floods, desertification, and drought brought on by climate change. In the North Atlantic, Category 5 hurricanes, the most destructive kind, occur three to four times more often than they did a decade ago.

While no single weather event can be tied directly to global warming, droughts, dust storms, and wildfires are becoming more common worldwide, and climate models predict that trend will accelerate. Southern California's worst wildfire in 30 years scorched 20,000 acres last spring [2009]. And in September [2009], Sydney, Australia, choked on its own version of the Dust Bowl: More than 5,000 tons of orange dirt swirled around the city during one of the region's worst droughts.

We're no longer talking about future generations; it's about us.

> ## FAST FACT
>
> A 2008 report published jointly by the Global Wind Energy Council and Greenpeace International found that using wind power to provide 12 percent of the world's energy needs would prevent as much as 1.5 billion tons of carbon dioxide from being released into the atmosphere each year.

Carbon Emissions Continue to Rise

Why haven't our leaders responded? They have been relying on old, conservative estimates of global warming effects. The 2007 Intergovernmental Panel on Climate Change (IPCC) projections used baseline scenarios from the 1990s, when scientists and government leaders assumed that by now, popular and political support would have led us to reduce greenhouse gas emissions. That means politicians and the people they represent have been looking at optimistic projections based on improvements that didn't happen.

In fact, global fossil-fuel and industrial carbon emissions have grown by 3.5 percent a year since 2000, faster than the worst-case scenario predicted by the Nobel Prize–winning IPCC. Atmospheric carbon dioxide concentrations are now at their highest levels in the last 15 million years, since before humans walked the earth.

U.N. senior official Luc Gnacadja recently told the press that by 2025, 70 percent of the world's land could be suffering from drought. In the United States, a report by the Union of Concerned Scientists says that in just a couple of decades, average summers in the country's bread basket, Illinois, could be hotter than the 1988 heat wave that wiped out $40 billion worth of food crops. In the next 12 years, there's a 50-50 chance that a combination of climate change and overuse will dry up Lakes Mead and Powell, say scientists with the Scripps Institution of Oceanography. Mead and Powell supply 90 percent of Las Vegas' water, along with irrigation and drinking water for more than 20 million people in Los Angeles and across Nevada and Arizona.

Not Too Late to Change—but Time Is Running Out

The vast majority of scientists agree that if we keep the Earth's temperature from rising 3.6 degrees Fahrenheit (2 degrees Celsius) above pre-industrial levels, we have a fighting chance of avoiding the most civilization-shaking impacts of climate change. The G8 leaders [of the eight nations with the top economies] agreed to that target at their July [2009] meeting.

Shoot past this limit, and the planet's ecosystems may enter a point of no return. We push the Earth into vicious spirals of feedback loops that make things even hotter. Sea ice melts, and the dark, open ocean absorbs more heat. The Amazon rainforest burns and releases even more greenhouse gases into the atmosphere. Weather patterns like El Niño transform from occasional to annual hurricane-brewing phenomena. Grain crops fail. One to 3 billion people face water shortage. The basic systems that support us, our societies, and life on the planet start breaking apart.

We Must Switch to Renewable Energy—and Fast

We have a choice to make. According to a consensus of hundreds of climate scientists, we can avert crashing the planet only if we make a sharp global U-turn by 2015: Level off emissions worldwide and bring them back down in the next few decades.

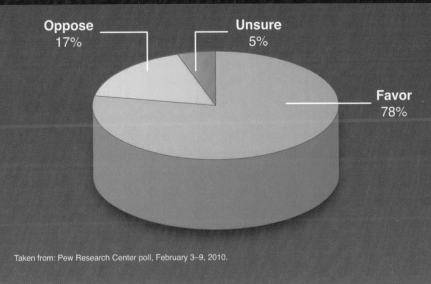

Americans Want Renewable Sources of Energy

A 2010 survey found that the majority of Americans want the government to invest in the research and development of renewable resources.

Question:
Would you favor or oppose the government increasing federal funding for research on wind, solar, and hydrogen technology?

Oppose
17%

Unsure
5%

Favor
78%

Taken from: Pew Research Center poll, February 3–9, 2010.

To do this, we must switch to much more efficient transportation, manufacturing, and buildings, and to solar, wind, tide, and biomass energy. Agriculture must make a rapid switch to organic and ecologically sound practices. The Worldwatch Institute estimates that livestock are responsible for more than half of greenhouse-gas emissions worldwide. We have to stop destroying forests for cattle ranches, palm oil plantations, and paper pulp, so we can preserve their ability to soak up carbon.

We need the world's governments to form ambitious and binding agreements at [the 2010 global conference on climate change in] Copenhagen and beyond. These agreements need to regulate and put a high price on emissions, and create incentives for a transition to a clean energy economy. The agreements must include help for the Global South [the Southern Hemisphere] in making the transition to a green economy.

Investing in Green Energy Is Affordable

We can afford to do this. *The Economics of 350*, recently released by the Economics for Equity and the Environment Network, says the cost of reducing CO_2 to 350 parts per million—the amount necessary to avoid a 3.6-degree temperature rise—would be between 1 percent and 3 percent of world GDP [gross domestic product]. It will cost far less than the 3.3 percent of GDP spent globally on insurance or the 4 percent-plus of GDP the United States spends on its military. And it will do more than either of those to increase our security.

Investments in renewable energy, building retrofits, and efficient mass transit will put people to work and create whole new industries, kick-starting an economic recovery that immediately benefits ordinary people.

The low-carbon culture we need to prevent climate catastrophe is not a culture of deprivation. We can move away from consumption for consumption's sake, gaining time to enjoy our lives more fully, and creating a world where our children and grandchildren have the opportunity to thrive. . . .

There Are Many Ways to Act

People are energized by the prospect of a green economy and new, clean technologies, and they want the opportunity not only to avert a climate catastrophe, but to help build a better future. Seventy-seven percent of those polled by Public Agenda say "investing in creating ways to get energy from alternative sources like solar and wind" is the best way to get the economy going, while just 16 percent believe "investing in finding more sources of oil, coal, and natural gas" is the answer.

This excitement is especially evident among those who were left out of the last wave of economic growth and today are sidelined by the so-called "jobless recovery." Among the supporters of green jobs and clean energy are urban youth, steelworkers, solar developers, architects, farmers, and all sorts of people who see the prospects for a green economic recovery that actually puts people to work.

The climate movement is demanding action in Washington, but not waiting for Congress to act. Businesses are adopting green practices and walking out on climate-denier associations, like the U.S. Chamber of Commerce.

Workplaces, homes, places of worship, and schools are being upgraded to become more climate-friendly and less costly to operate. Communities are making serious commitments to reducing their carbon emissions, restructuring their economies, and making neighborhoods green, resilient, and inclusive.

Buy-local campaigns are cutting down on long-distance transport and climate emissions. The strengthened local economies offer diverse livelihoods that meet people's immediate needs while weaving together the relationships that help people weather anything from an economic downturn to climate catastrophe. . . .

Time for a Green Revolution

We can still avert the extreme droughts, floods, storms, and displacements that could result if climate change reaches critical tipping points. It's still possible to save ourselves and generations to come from a climate so unstable that it can no longer support civilization as we know it. But we can't leave it to our leaders to fix it; the possibility only exists if we rise up and act now.

EVALUATING THE AUTHORS' ARGUMENTS:

The authors of this viewpoint claim that fossil fuel use contributes to climate change. But Megan Tady, author of the next viewpoint, claims that some energy alternatives contribute to climate change. In your opinion, to what extent can switching to energy alternatives help avert climate change? Explain your reasoning and quote from the texts you have read.

"Is there a word beyond irony to describe a plan to mitigate climate change that relies on cutting down the very trees that naturally remove carbon from the atmosphere? Stupidity, perhaps?"

Some Energy Alternatives Contribute to Climate Change

Megan Tady

Megan Tady is a blogger and video producer for Free Press, a national media reform organization. In the following viewpoint she argues that biofuel—fuel made by burning corn, palm oil, sugar cane, and other crops—is not a green renewable resource. In fact, argues Tady, biofuels actually contribute to global warming. Tady says it requires a lot of climate-change-causing energy to grow and harvest biofuels. In addition, Tady notes the production of biofuels requires forests to be cut down—but forests, which absorb carbon from the atmosphere, are a key ally in the fight against climate change. Tady says that biofuels add insult to injury by diverting food away from hungry people. For all of these reasons she concludes that biofuels are not a fix to the climate change problem.

Megan Tady, "Biofuels Are No Cure for Climate Change," *In These Times*, November 8, 2007. Reprinted by permission.

1. What does the phrase "insidious reality" mean in the context of the viewpoint?
2. What did an October 2007 study by Paul Crutzen find about the fertilizers used to grow biofuels? What bearing does this have on the author's argument?
3. What effect does the author say biofuels production might have on food prices?

S igh. Another day, another inane strategy to fight global warming. The bee in my bonnet this time is biofuels. They're nothing new, but governments and corporations are pushing biofuels with a renewed ferocity as the panacea for our ailing planet. But just as biofuels are working their way into our climate-cures lexicon, organizations, environmentalists and even the United Nations are blowing a very loud whistle. They warn that the United States and the European Union's renewable energy plans, which rely on biofuels, will have devastating impacts for the global South, turn our gaze away from investing in truly carbon-free technologies, and even add a flame to the fire igniting climate change.

Biofuels Are Not the Green Fuel They Appear to Be

Last month [October 2007], Jean Ziegler, a U.N. expert, called for a five-year moratorium on biofuel production, telling the Associated Press, "The effect of transforming hundreds and hundreds of thousands of tons of maize, of wheat, of beans, of palm oil, into agricultural fuel is absolutely catastrophic for the hungry people."

Last week, the humanitarian organization Oxfam International denounced the EU's proposal for 10 percent of transport fuels to come from biofuels by 2020, saying it could "spell disaster for some of the world's poorest people." The target for renewable-fuel use in the United States is 35 billion gallons a year.

We're being battered left and right with ominous news about climate change, so the idea of filling our tanks and heating our homes with biofuels is naturally comforting. Biofuels sound green. They're made from things that were once green—corn, palm oil, sugar cane and other agricultural products. And they're being touted as green. A Department of Energy's resource page for biofuels says, "Hey students!

Biofuels such as bioethanol and biodiesel can make a big difference in improving our environment."

But don't judge a climate cure by its color. Give it a rub, and you'll find that the term "biofuels" is actually obscuring an insidious reality. For that reason, many people, especially in the global South, have taken to calling them "agrofuels."

Biofuels Production Threatens the Planet

Consider this statement from the Landless Worker's Movement in Brazil in March [2007], where biofuel production is skyrocketing: "We can't call this a 'bio-fuels program.' We certainly can't call it a 'bio-diesel program.' Such phrases use the prefix 'bio' to subtly imply that the energy in question comes from 'life' in general. This is illegitimate and manipulative. We need to find a term in every language that describes the situation more accurately, a term like agro-fuel. This term refers specifically to energy created from plant products grown through agriculture."

And it's this agricultural production that has so many people worried. Biofuels need land, which means traditional food crops are being elbowed off of the field for fuel crops. Biofuel production is literally taking the food out of people's mouths and putting it into our gas tanks. Already, increased food costs sparked by increased demand are leaving populations hungry. The price of wheat has stretched to a 10-year high, while the price of maize has doubled.

Need more land? Clear cut some forest. Is there a word beyond irony to describe a plan to mitigate climate change that relies on cutting down the very trees that naturally remove carbon from the atmosphere? Stupidity, perhaps? The logic is like harvesting a sick patient's lungs to save her heart. Huge tracks of Amazon rainforest are being raised to the biofuels altar like a sacrificial lamb, and the UN suggests that 98 percent of Indonesia's rainforest will disappear by 2022, where heavy biofuel production is underway.

Biofuels Cause Climate Change

Still need land? Just take it. The human rights group Madre, which is backing the five-year moratorium, says agrofuel plantations in Brazil and Southeast Asia are displacing indigenous people. In an editorial published on CommonDreams last week, Madre Communication Director

Yifat Susskind wrote, "People are being forced to give up their land, way of life, and food self-sufficiency to grow fuel crops for export."

If this climate cure had a prescription bottle, the side effects would read: "Biofuels may cause drowsiness, headaches, human rights abuses, land deforestation, water depletion, worldwide hunger, and climate change." Wait, climate change? That's right; this cure is actually a cause.

Biofuels themselves may have a small carbon footprint, but the energy used to grow and process the fuel make for one large bear paw in the mud. Biofuels depend on the manufacturing of fertilizers, fuel used to power equipment, and fuel used to transport crops and fuels, which can offset any gains made in using biofuels. An October [2007] study by the Nobel Laureate Paul Crutzen determined that usage of nitrogen fertilizers causes biofuels to contribute more to global warming than petrol.

The Department of Energy (DOE) says biomass products, of which biofuels are derived, are "often more environmentally benign than their petroleum-derived counterparts." If the DOE was a betting man, how much would it wager on 'often?'

Biofuels Threaten the Poor and Hungry

The movement against biofuels has grown from a groundswell to a tidal wave. In January, more than 220 organizations worldwide appealed to the European Parliament to abandon their mandatory biofuels target. Even the International Monetary Fund is feeling nervous. In October, an IMF research team posted an article on the IMF's website which noted, "Until new technologies are developed, using food to produce

This biofuel production facility makes fuel from burning corn, sugar cane, palm oil, and other crops.

How Green Are Biofuels?

Not all biofuels are good for the environment. Many require large amounts of water, energy, fertilizer, and pesticides. Others reduce the food supply or take up large swaths of land in their production. More environmentally-friendly biofuels are not yet ready for production.

FUEL SOURCES

Crop	Used to produce	Greenhouse gas emissions* Kilograms of carbon dioxide created per mega joule of energy produced	Use of resources during growing, harvesting, and refining of fuel				Percent of existing US cropland needed to produce enough fuel to meet half of US demand	Pros and cons
			Water	Fertilizer	Pesticide	Energy		
Corn	Ethanol	81-85	high	high	high	high	157%-262%	Technology ready and relatively cheap, reduces food supply
Sugar-cane	Ethanol	4-12	high	high	med	med	46%-57%	Technology ready, limited as to where will grow
Switch grass	Ethanol	-24	med-low	low	low	low	60%-108%	Won't compete with food crops, technology not ready
Wood residue	Ethanol, biodiesel	N/A	med	low	low	low	150%-250%	Uses timber waste and other debris, technology not fully ready
Soybeans	Biodiesel	49	high	med-low	med	med-low	180%-240%	Technology ready, reduces food supply
Rapeseed, canola	Biodiesel	37	high	med	med	med-low	30%	Technology ready, reduces food supply
Algae	Biodiesel	-183	med	low	low	high	1%-2%	Potential for huge production levels, technology not ready

*Emissions produced during the growing, harvesting, refining, and burning of fuel. Gasoline is 94, diesel is 83.

Taken from: Lisa Stiffler, "Bio-debatable: Food vs. Fuel," *(Seattle) Post-Intelligencer*, May 3, 2008. www.seattlepi.com/dayart/20080503/biofuels_compare.gif.

biofuels might further strain already tight supplies of arable land and water all over the world, thereby pushing food prices up even further."

It's new, carbon-free and sustainable technologies that we need to be investing in, rather than a plan that has as much stock as [US president

George W.] Bush's missile defense scheme. Madre said that the moratorium on biofuels should be accompanied by technologies that don't compromise global food security.

Susskind wrote, "We need sustainable solutions to climate change, not corporate solutions that seek to simply shift our energy addiction from one resource to another. Creative and practical solutions for meeting our energy requirement—including some local, sustainable biofuel programs—are being developed around the world." In theory, biofuel production could reduce poverty by increasing jobs for small farmers around the world. But Oxfam warns that the "huge plantations emerging to supply the EU pose more threats than opportunities for poor people." If we're going to pull the current form of biofuels production out from under places like debt-riddled Brazil, we need to replace it with another plan that offers sustainable economic development for poor communities.

Biofuels Are Not the Answer

Oxfam suggests the EU implement safeguards in biofuel production that protect land rights, livelihoods, workers rights and food security. "The EU set its biofuel target without checking the impact on people and the environment," Oxfam spokesperson Robert Bailey said in a press release. "The EU must include safeguards to ensure that the rights and livelihoods of people in producing countries are protected. Without these, the ten per cent target should be scrapped and the EU should go back to the drawing board."

Until then, the Department of Energy is exactly right. Biofuels will make a big difference in improving our environment—"our" being the United States and the EU, and nobody else.

EVALUATING THE AUTHOR'S ARGUMENTS:

Megan Tady quotes from several sources to support the points she makes in her essay. Make a list of all the people she quotes, including their credentials and the nature of their comments. Then pick the quote you found most persuasive. Why did you choose it? What did it lend to Tady's argument?

Dependence on Fossil Fuels Fosters Geopolitical Instability

Daveed Gartenstein-Ross

"A major terrorist attack against the oil supply would dramatically change the global order."

In the following viewpoint Daveed Gartenstein-Ross argues that America's dependence on oil makes it vulnerable to terrorism and chaos. The author warns that terrorists such as Osama bin Laden have tried to target oil production facilities in Saudi Arabia and elsewhere. If they successfully took out a major facility, Gartenstein-Ross predicts America would be paralyzed. It would no longer have access to its most valuable energy source, and all industries that rely on oil—such as food delivery, transportation, and other vital sectors—would come to a standstill. Panic and astronomically high oil prices would quickly follow. For these reasons Gartenstein-Ross concludes that transitioning to alternative energy sources is best for America's security. Gartenstein-Ross is the vice president of research at the Foundation for Defense of Democracies. He is the editor of the book *From Energy Crisis to Energy Security: A Reader.*

Daveed Gartenstein-Ross, "Jihad for Oil," *Weekly Standard*, August 14, 2008. Reprinted by permission.

AS YOU READ, CONSIDER THE FOLLOWING QUESTIONS:
  1. What did Osama bin Laden say in a December 2004 audiotape about how to defeat Western countries?
  2. Who is Robert Baer, and how does he factor into the author's argument?
  3. What would happen if terrorists successfully attacked the Abqaiq facility, according to the author?

Oil dependence is America's Achilles' heel in the battle against terrorism—a fact that has not escaped the terrorists. Osama bin Laden and others have declared the oil supply a top target, and subsequent plots demonstrate that the desire to disrupt world energy markets is more than mere rhetoric. This significant weakness should factor heavily in current political debates about alternatives to oil.

Terrorists Target Oil Production

When bin Laden dramatically addressed the United States in a video released on the eve of the 2004 elections, he boasted of his "bleed-until-bankruptcy" plan for defeating America. His focus on the economy is a primary reason that the terrorist leader reversed his original pledge to keep oil off limits as a military target. In his 1996 declaration of war against America, bin Laden said that oil was not part of the battle because it was "a large economical power essential for the soon to be established Islamic state," but in a December 2004 audiotape he reversed this promise. Declaring Western countries' purchase of oil at then-market prices "the greatest theft in history," he stated: "Focus your operations on it [oil production], especially in Iraq and the Gulf area, since this [lack of oil] will cause them to die off [on their own]."

Bin Laden's deputy Ayman al-Zawahiri called for al-Qaeda fighters to "concentrate their campaigns on the stolen oil of the Muslims" in a December 2005 video. Likewise, *Sawt al-Jihad*, the online magazine of al-Qaeda in the Arabian Peninsula, claimed in February 2007 that cutting the U.S.'s oil supplies "would contribute to the ending of the American occupation of Iraq and Afghanistan."

Saudi Arabia is the most critical oil-producing country that terrorists have targeted. Saudi efforts are vital to stability of the worldwide oil supply because that country holds 25 percent of the globe's proven reserves, produces almost 10 million barrels per day, and is the only country that can maintain excess production capacity of around 1.5 million barrels per day (a "swing reserve") to keep world prices stable. However, Saudi production is particularly vulnerable to attack because it depends on a limited number of hubs. Two-thirds of Saudi Arabia's oil is processed at the Abqaiq facility, and there are two main export terminals: Ras Tanura and Ras al-Ju'aymah.

Terrorists have in fact directed their efforts toward attacks against these hubs. In September 2005, following a 48-hour shootout with a cell in the seaport of al-Dammam, police discovered forged documents that would have given the terrorists access to some of Saudi Arabia's key oil and gas facilities.

Oil Dependence Can Bring America to Its Knees

Terrorists affiliated with al-Qaeda in the Arabian Peninsula also obtained sensitive access to facilities for a February 2006 attack on the Abqaiq refinery, which is operated by the state-owned Saudi Aramco. Though local news sources played down the attack (one even described it as proof of "how tight and impenetrable the existing Saudi security system is"), written evidence submitted to Britain's House of Commons by Neil Partrick, a senior analyst in The Economist Group's Economist Intelligence Unit, paints a different picture. Noting that the attackers wore Aramco uniforms, drove Aramco vehicles, and were able to enter the facility's first perimeter fence, Partrick concludes that either the terrorists "had inside assistance from members of the formal security operation of the state-owned energy company" or else security was so lax "that these items could be obtained and entry to the site obtained." Either possibility is a concern.

Could a catastrophic attack against Saudi production succeed? Such an attack could be executed using tactics that al-Qaeda has successfully employed in the past. For example, it would be difficult to safeguard facilities against an airplane used as a guided missile, à la 9/11. Thus, former CIA case officer Robert Baer wrote in his 2003 book *Sleeping with the Devil:* "A single jumbo jet with a suicide bomber at the controls, hijacked during takeoff from Dubai and crashed into the heart of Ras Tanura, would be enough to bring the world's oil-addicted economies to their knees, America's along with them."

The Abqaiq refinery that was targeted in February 2006 is also a critical point of vulnerability. If a major attack is successfully executed in one of these locations, the reduced worldwide oil supply would be joined by an inflated risk premium. Julian Lee, a senior energy analyst at the Centre for Global Energy Studies in London, told the *Guardian* in 2004 that following a significant loss of Saudi oil, "it would be difficult to put an upper limit on the kind of panic reaction you would see in the global oil markets." The ramifications would be not only economic but also military: Sawt al-Jihad may be correct that such an attack could doom U.S. ventures in Afghanistan and Iraq.

A Threat to the Global Order

In addition to catastrophic attacks, terrorists can undertake disruptive attacks against specific nodes. The recent activities of the Movement for the Emancipation of the Niger Delta (MEND) show the effect that disruptive attacks can have. Saudi Arabia pledged to produce an extra 200,000 barrels of oil per day [bpd] beginning in July 2008 to curb record prices, yet MEND and its copycats knocked more than that offline in a single week: an attack on Shell's Bonga field

Iraqi police patrol an oil pipeline in Iraq. Experts fear a major attack on oil production facilities and the subsequent detrimental effects on the world's economy.

coupled with two attacks on Chevron's Abiteve Olero crude oil line cut Nigeria's output by about 400,000 bpd. Though the Nigerian facilities will be repaired, this demonstrates how disruptive attacks can scotch the market's supply expectations.

The situation is growing more rather than less perilous: Gal Luft and Anne Korin of the Institute for the Analysis of Global Security have noted that growing worldwide demand has reduced OPEC's spare capacity from seven million barrels a day in 2002 to only two million today (less than 2.5 percent of the market). "As a result," they write, "the oil market today resembles a car without shock absorbers: the tiniest bump can send a passenger to the ceiling." Moreover, global consumption is only expected to increase: the world is projected to have 1.25 billion cars on the road in 2030, up from 700 million today.

A major terrorist attack against the oil supply would dramatically change the global order, in ways that most policymakers have probably never contemplated. The threat of terrorism thus adds urgency to current discussions about alternatives to oil.

EVALUATING THE AUTHOR'S ARGUMENTS:

Daveed Gartenstein-Ross hinges his argument that fossil fuel dependence creates instability on the fact that so much of the US economy is powered by oil. List at least five items in your house or classroom that are either made from, powered by, or produced as a result of fossil fuels. Then, consider what might happen if oil were either unavailable or too expensive to buy. Do you agree with Gartenstein-Ross that panic would ensue? Why or why not?

Transitioning to Alternative Energies Will Foster Geopolitical Instability

"Greening the world will certainly eliminate some of the most serious risks we face, but it will also create new ones."

David Rothkopf

In the following viewpoint David Rothkopf argues that alternative energy sources will herald their own conflicts, shortages, and threats. Rothkopf says that fighting and instability over oil is well known, and one of the reasons many want to switch to alternative energies. Yet he warns that such a switch will not be without turmoil. Alternatives to fossil fuels include nuclear power, biofuels, and lithium batteries, but Rothkopf says each of these sources stirs up at least one political, environmental, or military threat or challenge. Rothkopf suggests that the United States invest in alternative energies but be aware that a greener world will not be devoid of risk, challenge, or threat. Rothkopf is president and chief executive of Garten Rothkopf, a Washington-based advisory firm specializing in energy, climate,

David Rothkopf, "Is a Green World a Safer World? A Guide to the Coming Green Geopolitical Crises," *Foreign Policy,* September–October 2009. Reprinted by permission.

and global-risk-related issues. He is a visiting scholar at the Carnegie Endowment for International Peace and author of *Superclass: The Global Power Elite and the World They Are Making.*

AS YOU READ, CONSIDER THE FOLLOWING QUESTIONS:
1. Name at least three alternative energy sources or machines that Rothkopf says rely heavily on water.
2. What types of machines do lithium-ion batteries power, according to Rothkopf? List at least three.
3. What could ignite conflict between Bolivia and Chile, according to Rothkopf?

Greening the world will certainly eliminate some of the most serious risks we face, but it will also create new ones. A move to electric cars, for example, could set off a competition for lithium, another limited, geographically concentrated resource. The sheer amount of water needed to create some kinds of alternative energy could suck certain regions dry, upping the odds of resource-based conflict. And as the world builds scores more emissions-free nuclear power plants, the risk that terrorists get their hands on dangerous atomic materials—or that states launch nuclear-weapons programs—goes up.

The decades-long oil wars might be coming to an end as black gold says its long, long goodbye, but there will be new types of conflicts, controversies, and unwelcome surprises in our future (including perhaps a last wave of oil wars as some of the more fragile petrocracies decline). If anything, a look over the horizon suggests the instability produced by this massive and much-needed energy transition will force us to grapple with new forms of upheaval. . . .

Nuclear Power, Nuclear Threat

There is simply no way to reverse the effects of climate change without much more broadly embracing nuclear energy. Not only is it essentially emissions free, scalable, and comparatively energy efficient, but just 1 metric ton of uranium produces the same amount of energy as approximately 3,600 metric tons of oil (about 80,000 barrels). It is a far more sophisticated and proven technology than virtually all of the other emerging alternatives. These facts have already led to a very real renaissance in nuclear energy, one that is concentrated in the energy-hungry developing world (more than two thirds of announced projects are in developing countries).

Unfortunately, nuclear power is also fraught with real and perceived risks. Plant-safety hazards are pretty minimal, if history is any indicator. However, two real issues loom. One is how to safely dispose of spent fuel, a dilemma still hotly debated by environmentalists. And another is how to ensure the security of the fuel at every other stage of its life cycle, particularly in comparatively cash-strapped emerging countries, which are often in regions scarred by instability and home to terrorist organizations with their own nuclear ambitions.

With each new program, the chances of a security breach increase. Nor is the danger of a bad actor diverting fuel to produce an atomic bomb the only nuclear nightmare we're facing. Radioactive waste could be used to produce a dirty bomb with devastating impact. And fiddling with weapons programs behind closed doors might be the greatest security risk of all.

Nuclear-weapons expert Robert Gallucci once told me that, considering these growing risks, a deadly nuclear terrorist incident was "almost certain." Such an event would have broad global aftershocks affecting areas as diverse as civil liberties and trade. Imagine, for example, trying to ship anything anywhere in the world the day after. To give just one example, only 5 percent of shipping containers today are subject to visual inspection in the United States. Pressure to make inspection absolute in the wake of a nuclear event could easily lead to the buildup of millions of goods at U.S. ports, driving up consumer-goods prices as market supplies dwindle.

A new nuclear nonproliferation treaty is already on the drawing board, but even as U.S. President Barack Obama works to fulfill his dream of a world free of nuclear weapons, it is already clear that the risks posed by old-fashioned national stockpiles are being eclipsed by those associated with small groups exploiting cracks in an increasingly complex worldwide nuclear infrastructure.

Water as "The New Oil"

Today, 1.1 billion people don't have ready access to clean water, and estimates suggest that within two decades as many as two thirds of

Containers of brine to extract lithium are seen at the Uyuni salt lake in Bolivia. Three quarters of the world's lithium deposits are concentrated in an area that could cause conflict between Bolivia, Argentina, and Chile.

the Earth's people will live in water-stressed regions. It has become a new conventional wisdom that water will become "the new oil," as Dow Chemical Chief Executive Andrew Liveris has said, both because of the new value it will have and the new conflicts it will generate.

Ironically, the hunt for energy alternatives to replace oil could make the water problem much worse. Some biofuels use significant amounts of water, including otherwise efficient sugar cane (unlike rain-soaked ethanol giant Brazil, most sugar-cane producers have to irrigate). Similarly, the various technologies that are seen as essential to the clean use of coal are water hogs. Plug-in hybrid cars also increase water use because they draw electricity, and most types of power plants use water as a coolant. Even seemingly unrelated technologies, such as silicon chips (key to everything from smart-grid technologies to more efficient energy use) require a great deal of water to produce.

Many countries could begin to address this by working out schemes to charge for water, the single best way to grapple with this problem. Alternatively, they may build nuclear desalination plants that make saltwater drinkable. Neither course is perfect. A de facto privatization of water has occurred throughout the world, with low-income populations forced to purchase bottled water to avoid contamination, but even so, the ideal of the right to free water has held firm and governments have found it politically untenable to charge even nominal sums. And those nuclear desalination plants? As countries that have deployed this technology, such as India, Japan, and Kazakhstan, have found, they're bloody expensive, at hundreds of millions of dollars a pop.

The Great Lithium Game

In Asia, Europe, and the United States, people are getting excited about the electric car—and for good reason. Electric cars will enable greater independence from oil and could play a significant role in lowering carbon dioxide emissions. But the major fly in the ointment for the electric car is the battery.

Many solutions are being considered, including "air" batteries that produce electricity from the direct reaction of lithium metal with oxygen. The most likely option for now, though, is the lithium-ion battery used in cameras, computers, and cellphones. Lithium-ion batteries offer better storage and longer life than the older nickel-metal hydride models, making them ideal for a space-constrained, long-running vehicle.

All this means that lithium is likely to be a hot commodity in the years immediately ahead. It so happens that about three quarters of the world's known lithium reserves are concentrated in the southern cone of Latin America—to be precise, in the Atacama Desert, which is shared by two countries: Chile and Bolivia. Other than these reserves and the Spanish language, the one thing these two countries have in common is a historical animosity, cemented by their late 19th-century War of the Pacific. Chile was able to cut off Bolivia's access to the sea, a maneuver that rankles bitterly in La Paz to this day.

FAST FACT

Lithium, the mineral used to make lithium-ion batteries that fuel next-generation electronics and electronic vehicles, is a finite resource that must be mined. Lithium ores are located in just a few areas of the world.

Bolivia's lack of coastline could become an issue again if the two lithium powerhouses start jostling to attract investors. Competition between Bolivian and Chilean lithium mines and, potentially, over domestic production of lithium batteries could very well bring about a second War of the Pacific—to say nothing of the huge environmental costs that lithium mining incurs. Any such tension could jeopardize U.S. efforts to adopt electric vehicles, as the United States already gets 61 percent of its lithium imports from Chile. China and Russia, which also hold significant reserves, would be poised to ride out and profit from such an event. Further, conflict between the two Latin American states would likely bolster the fortunes of batteries made from less efficient resources, such as those used in nickel-metal hydride batteries, or boost other technologies that use different substances with their own drawbacks. And in any event, the possibility of a regional lithium rush reminds us that whatever technologies take hold, demand will emerge for the scarce commodities on which they depend . . . and we know well where that can lead.

The United States Should Proceed with Caution

These are just a few, fleeting glimpses of the future, but many geopolitical ramifications of moving toward green energy are very much with us already. . . .

The bottom line: A shift away from dirty old fuels is the only path toward reducing several of the greatest security threats the planet faces, but we must step carefully and avoid letting our optimism run away with us. By acknowledging that a greener world will hardly be devoid of geopolitical challenges and preparing accordingly, we may find a path to defusing our threats today, while largely avoiding the inadvertent drawbacks of desperately needed innovation.

EVALUATING THE AUTHOR'S ARGUMENTS:

In the viewpoint you just read, Rothkopf describes the tensions, conflicts, and shortages that are likely to result from transitioning to alternative energy sources. Given what you know on this topic, what suggestions would you offer for easing the tensions and shortages Rothkopf predicts will occur? How might the United States be able to ensure a smoother transition to energy alternatives? Write at least two paragraphs in which you detail at least one suggestion.

Chapter 2

Are Alternative Sources of Energy Viable?

The two most popular alternative energy sources are wind and solar power. But their overall viability remains controversial.

Renewable Energy Can Meet Demand in the United States

John Farrell and David Morris

"All 36 states with either renewable energy goals or renewable energy mandates could meet them by relying on in-state renewable fuels."

In the following viewpoint David Morris and John Farrell argue that renewable energy sources like solar and wind power can meet demand for energy in the United States. They claim that solar power panels are particularly well suited for this task: Putting such panels on rooftops could generate huge amounts of electricity that could help states avoid burning fossil fuels and keep energy costs down. Morris and Farrell say the key to using renewable energy successfully in the United States is to introduce innovative technologies on a local level. Small-scale renewable technologies can help states meet or even exceed their renewable energy goals. Morris and Farrell conclude that states have the power to become energy self-reliant by incorporating renewable resource technologies into their energy grid. David Morris is cofounder and vice president of the Institute for Local Self-Reliance, where John Farrell is a senior researcher.

John Farrell and David Morris, "Energy Self-Reliant States: Executive Summary," Institute for Local Self-Reliance (ILSR), May 2010. Reprinted by permission.

AS YOU READ, CONSIDER THE FOLLOWING QUESTIONS:
  1. What percent of states with renewable energy goals do Morris and Farrell say could supply 100 percent of their electricity from in-state renewable sources?
  2. How many states do the authors claim could generate 25 percent of their electricity from rooftop photovoltaic panels alone?
  3. How much less expensive is solar electricity in Nevada than in Iowa, according to the authors? How much less than in Pennsylvania?

I n 2009, the nation is involved in a vigorous and far reaching debate about the scale of future energy systems. As we shift from fossil fuels to renewable energy a new question looms before us. Will we embrace a centralized renewable energy future characterized by greater federal involvement in planning, or will we meet local and state needs with local and state-based strategies? . . .

States Are Capable of Generating Their Own Power

All 36 states with either renewable energy goals or renewable energy mandates could meet them by relying on in-state renewable fuels. Sixty-four percent could be self-sufficient in electricity from in-state renewables; another 14 percent could generate 75 percent of their electricity from homegrown fuels.

Indeed, the nation may be able to achieve a significant degree of energy independence by harnessing the most decentralized of all renewable resources: solar energy. More than 40 states plus the District of Columbia could generate 25 percent of their electricity just with rooftop PV [photovoltaics].

In fact, these data may be conservative. [This] report does not, for example, estimate the potential for ground photovoltaic arrays—although it does estimate the amount of land needed in each state to be self-sufficient relying on solar—even though common sense suggests that this should dwarf the rooftop potential.

Local Level Technology Makes the Difference

Even as FERC [the Federal Energy Regulatory Commission] and Congress and environmental groups, spurred by independent

Alternative energy sources are shown installed on a house. The wind turbine generates electricity from the wind, photovoltaic solar panels harness the energy in sunlight, and solar heat panels (center on top row) heat water to be used as heat or hot water.

renewable power producers (some of the biggest of whom are sub-sidiaries of regulated utilities) rush to pre-empt state authority and accelerate the construction of a new $100–200 billion interregional transmission network, the case for state-focused planning has never been stronger.

For it is at the state, not the federal level, that comprehensive, least cost energy planning is used. It is at the state—not the federal or multi-state regional level—that efficiency, demand reduction, distributed generation and other commercially available strategies are often evaluated together.

It is at the local level that new technologies like smart grids, electric vehicles, distributed storage, and rooftop solar will have their major impact. The integration of millions of electric vehicles into the grid, for example, will change the context for energy planning by creating, for the first time, abundant storage for electricity.

It is at the state and local level that the most important new energy developments are taking place. Efficiency Vermont has empirically proven that an aggressive electric conservation program can reduce current consumption even with economic and population growth. Cities like Berkeley [California] already are mapping their rooftop solar potential, installing charging stations for electric vehicles, and directly financing efficiency and renewable energy in households and businesses.

> # FAST FACT
>
> A 2010 poll conducted jointly by the *National Journal's Congressional Connection*, the Pew Research Center, and the Society for Human Resource Management reported that 78 percent of Americans favor legislation that would require utilities to produce more energy from wind, solar, and other renewable sources of energy.

Create Jobs and Save Money

Perhaps the most important reason to make states the principal actors in energy planning is that their collective economic self-interest is consistent with the national interest. Every state could create thousands of new jobs and hundreds of millions, perhaps billions, of dollars in economic development, through a vigorous strategy of energy efficiency and renewable energy.

Those promoting a new inter-regional transmission network argue that even if renewable energy is to be found everywhere, states with more reliable and higher speed winds or with more abundant sunshine can generate electricity cheaper.

That is undeniable. Nevada can produce solar electricity from photovoltaic panels at a price about 20 percent less than Iowa and about

35 percent less than Pennsylvania. A typical North Dakota commercial wind turbine can produce electricity at a cost about 30 percent less than one in Ohio.

But in most cases these significant variations result in modest variations in the retail cost of energy when the cost of transporting the energy is taken into account.

For example, if Ohio's electricity came from North Dakota wind farms—1,000 miles away—the cost of constructing new transmission lines to carry that power and the electricity losses during transmission could result in an electricity cost to the customer that is about the same, or higher, than local generation with minimal transmission upgrades.

Thus centralized renewable energy might not be in the nation's economic interest, even when the cost-benefit analysis focuses solely on the impact on the retail price. But if we were to use a more expansive definition of economic interest, that is, the impact of renewable energy development on local and state jobs and economies, state-based energy self-reliance strategies can be clear economic winners.

States Want Renewable Energy

States have clearly indicated their desire to harness renewable energy within their borders. For example, Ohio requires half of its renewable energy mandate to be met with in-state production. Colorado and Missouri each have a 1.25 multiplier for in-state resources used to meet their renewable energy requirements. Minnesota's Community-Based Energy Development statute encourages more locally owned wind power. Washington State offers solar incentive payments based on the portion of the panels made in the state, as well as reserving incentives for community solar.

The data . . . argue that a new extra high voltage inter-regional transmission network may not be needed to improve network reliability, relieve congestion and expand renewable energy. The focus should be on upgrading the transmission, subtransmission and distribution systems inside states. . . .

Many states have sufficient renewable energy to generate 100 percent of their electricity. Clearly this is a theoretical statement in the sense that it is very long term and to achieve high penetration rates

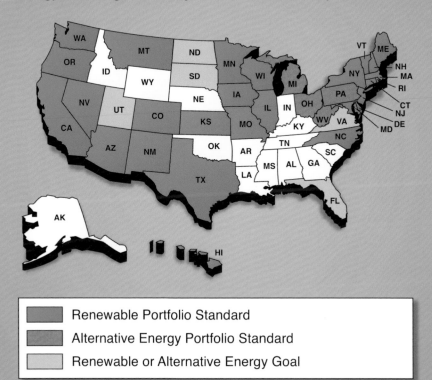

States Are Moving Toward Renewable Energy

The majority of US states have set renewable energy goals or standards. These state that by a certain year (such as 2020), a certain percentage (such as 15 or 20 percent) of the state's total energy must be generated by renewable sources of power.

Renewable Portfolio Standard

Alternative Energy Portfolio Standard

Renewable or Alternative Energy Goal

Taken from: Pew Center on Global Climate Change, December 2009.

of variable renewable energy will require significant developments in storage technology as well as significant investments in upgrading distribution and transmission networks to allow for massive amounts of dispersed generation.

Energy Self-Reliance Is Possible

Yet even if there is much to do before very high proportions of our electricity system can be generated by renewable energy, these data do suggest that for the foreseeable future states can and should harness homegrown renewable fuels to meet in-state demand. No state has yet exceeded 10 percent renewable electricity (excluding large scale hydro) and . . . all states with renewable energy mandates, no matter how high, could satisfy them by relying solely on in-state energy sources.

The power of states to control their renewable energy future hangs in the balance. Others have documented how states have used their authority to improve the prospects for renewable energy, from policies favoring domestic generation to smart grids and conservation programs. [There is] compelling evidence that if states retain their authority, energy self-reliance is within their grasp.

EVALUATING THE AUTHORS' ARGUMENTS:

In the viewpoint you just read, the authors use facts, statistics, examples, and reasoning to make their argument that renewable energy can meet demand in the United States. They do not, however, use any quotations to support their point. If you were to rewrite this article and insert quotations, what authorities might you quote from? Where would you place them, and why?

Renewable Energy Cannot Meet Demand in the United States

Robert Bryce

> *"[Even] by doubling wind and solar output by 2012, the contribution of those two sources to America's overall energy needs will still be almost inconsequential."*

Robert Bryce is the managing editor of *Energy Tribune* and the author of *Gusher of Lies: The Dangerous Delusions of "Energy Independence."* In the following viewpoint he argues that renewable sources of energy are not powerful enough to meet the demand for energy in the United States. Although solar and wind power use has nearly doubled in recent years, Bryce says power from these renewables continues to produce a minuscule portion of the country's total electricity consumption. Bryce concludes that fossil fuels—oil, natural gas, and coal—and nuclear power are the only energy sources that can offer the tremendous energy output required to power American society. He urges America's leaders to be honest about that fact as they craft the nation's energy policy.

Robert Bryce, "Let's Get Real About Renewables," *Wall Street Journal,* March 5, 2009. Reprinted by permission.

During his address to Congress last week [on February 24, 2009], President Barack Obama declared, "We will double this nation's supply of renewable energy in the next three years."

While that statement—along with his pledge to impose a "cap on carbon pollution"—drew applause, let's slow down for a moment and get realistic about this country's energy future. Consider two factors that are too-often overlooked: [former president] George W. Bush's record on renewables, and the problem of scale.

The Recurring Problem with Renewables

By promising to double our supply of renewables, Mr. Obama is only trying to keep pace with his predecessor. Yes, that's right: From 2005 to 2007, the former Texas oil man oversaw a near-doubling of the electrical output from solar and wind power. And between 2007 and 2008, output from those sources grew by another 30%.

Mr. Bush's record aside, the key problem facing Mr. Obama, and anyone else advocating a rapid transition away from the hydrocarbons that have dominated the world's energy mix since the dawn of the Industrial Age, is the same issue that dogs every alternative energy idea: scale.

Let's start by deciphering exactly what Mr. Obama includes in his definition of "renewable" energy. If he's including hydropower, which now provides about 2.4% of America's total primary energy needs, then the president clearly has no concept of what he is promising. Hydro now provides more than 16 times as much energy as wind and solar power combined. Yet more dams are being dismantled than built. Since 1999, more than 200 dams in the U.S. have been removed.

If Mr. Obama is only counting wind power and solar power as renewables, then his promise is clearly doable. But the unfortunate truth is that even if he matches Mr. Bush's effort by doubling wind and solar output by 2012, the contribution of those two sources to America s overall energy needs will still be almost inconsequential.

A Mere Fraction of the Nation's Energy

Here's why. The latest data from the U.S. Energy Information Administration show that total solar and wind output for 2008

Renewables Are a Small Part of Americans' Energy Consumption

US Energy Consumption by Energy Source

Fossil Fuels	83.531%
Coal	22.398%
Coal Coke Net Imports	0.040%
Natural Gas	23.814%
Petroleum	37.279%
Electricity Net Imports	0.113%
Nuclear Electric Power	8.427%
Renewable Energy	7.367%
Biomass	3.852%
Geothermal Energy	0.360%
Hydroelectric Conventional	2.512%
Solar Thermal/PV Energy	0.097%
Wind Energy	0.546%

Total: 99.438%

Note: Total does not equal sum of components due to rounding.

Taken from: US Energy Information Administration, March 2010.

will likely be about 45,493,000 megawatt-hours. That sounds significant until you consider this number: 4,118,198,000 megawatt-hours. That's the total amount of electricity generated during the rolling 12-month period that ended last November. Solar and wind, in other words, produce about 1.1% of America's total electricity consumption.

Of course, you might respond that renewables need to start somewhere. True enough—and to be clear, I'm not opposed to renewables. I have solar panels on the roof of my house here in Texas that generate 3,200 watts. And those panels (which were heavily subsidized by Austin Energy, the city-owned utility) provide about one-third of the electricity my family of five consumes. Better still, solar panel producers like First Solar Inc. are lowering the cost of solar cells. On the day of Mr. Obama's speech, the company announced that it is now producing solar cells for $0.98 per watt, thereby breaking the important $1-per-watt price barrier.

And yet, while price reductions are important, the wind is intermittent, and so are sunny days. That means they cannot provide the baseload power, i.e., the amount of electricity required to meet minimum demand, that Americans want.

That issue aside, the scale problem persists. For the sake of convenience, let's convert the energy produced by U.S. wind and solar installations into oil equivalents.

The conversion of electricity into oil terms is straightforward: one barrel of oil contains the energy equivalent of 1.64 megawatt-hours of electricity. Thus, 45,493,000 megawatt-hours divided by 1.64 megawatt-hours per barrel of oil equals 27.7 million barrels of oil equivalent from solar and wind for all of 2008.

Now divide that 27.7 million barrels by 365 days and you find that solar and wind sources are providing the equivalent of 76,000 barrels

The Bonneville Dam on the Columbia River generates hydroelectric power. Hydropower produces sixteen times as much energy as wind and solar combined, but it still only provides about 2.4 percent of America's total primary energy needs.

of oil per day. America's total primary energy use is about 47.4 million barrels of oil equivalent per day.

We Cannot Ignore the Power Offered by Fossil Fuels

Of that 47.4 million barrels of oil equivalent, oil itself has the biggest share—we consume about 19 million barrels per day. Natural gas is the second-biggest contributor, supplying the equivalent of 11.9 million barrels of oil, while coal provides the equivalent of 11.5 million barrels of oil per day. The balance comes from nuclear power (about 3.8 million barrels per day), and hydropower (about 1.1 million barrels), with smaller contributions coming from wind, solar, geothermal, wood waste, and other sources.

Here's another way to consider the 76,000 barrels of oil equivalent per day that come from solar and wind: It's approximately equal to the raw energy output of one average-sized coal mine.

During his address to Congress, Mr. Obama did not mention coal—the fuel that provides nearly a quarter of total primary energy and about half of America's electricity—except to say that the U.S. should develop "clean coal." He didn't mention nuclear power, only "nuclear proliferation," even though nuclear power is likely the best long-term solution to policy makers' desire to cut U.S. carbon emissions. He didn't mention natural gas, even though it provides about 25% of America's total primary energy needs. Furthermore, the U.S. has huge quantities of gas, and it's the only fuel source that can provide the stand-by generation capacity needed for wind and solar installations. Finally, he didn't mention oil, the backbone fuel of the world transportation sector, except to say that the U.S. imports too much of it.

Renewables Will Never Do What Oil Can

Perhaps the president's omissions are understandable. America has an intense love-hate relationship with hydrocarbons in general, and with coal and oil in particular. And with increasing political pressure to cut carbon-dioxide emissions, that love-hate relationship has only gotten more complicated.

But the problem of scale means that these hydrocarbons just won't go away. Sure, Mr. Obama can double the output from solar and wind. And then double it again. And again. And again. But getting from 76,000 barrels of oil equivalent per day to something close to the 47.4 million barrels of oil equivalent per day needed to keep the U.S. economy running is going to take a long, long time. It would be refreshing if the president or perhaps a few of the Democrats on Capitol Hill would admit that fact.

EVALUATING THE AUTHORS' ARGUMENTS:

Robert Bryce disagrees with David Morris and John Farrell (authors of the previous viewpoint) on whether renewable resources can generate significant quantities of energy that can power the United States. After reading both viewpoints, who do you think is right? What pieces of evidence swayed you? List at least two in your answer.

Wind Power Is a Viable Alternative Energy Source

Robert Thresher

"The U.S. Department of Energy (DOE) has worked ... to turn yesterday's dream for a clean, renewable energy source into today's most viable renewable energy technology."

In the following viewpoint, Robert Thresher discusses how wind-generated energy has improved over the last thirty years. Wind plants or wind farms can generate from 750,000 to 1,500,000 kilowatts, while small, single turbines can produce electricity on the scale of 100 kilowatts. The cost of wind power has come down from eighty cents per kilowatt hour to between four and six cents per kilowatt hour, and efforts are being made to bring the cost down further. In Europe, wind farms have been set up offshore in shallow water, and offshore wind farms are being designed for the deeper waters off the coast of the United States. These projected offshore wind farms are expected to help power desalination efforts to provide drinkable water for the world's growing population. Robert Thresher was director of the National Wind Technology Center at the US Department of Energy's National Renewable Energy Laboratory (NREL) from 1994 to 2008.

Robert Thresher, "Wind Is Most Viable Energy Renewable, Energy Department Says," America.gov, July 6, 2005.

AS YOU READ, CONSIDER THE FOLLOWING QUESTIONS:
1. What does the author describe as a low-wind speed site?
2. How much had wind energy production worldwide increased between 1994 and 2004, according to the author?
3. What does the author think the difference in design will be between European offshore wind turbines and those expected to be built to meet America's offshore wind energy needs?

The U.S. Department of Energy (DOE) has worked with the U.S. wind energy industry for more than 30 years to turn yesterday's dream for a clean, renewable energy source into today's most viable renewable energy technology.

Wind power—the technology of using wind to generate electricity—is the fastest-growing new source of electricity worldwide. Wind energy is produced by massive three-bladed wind turbines that sit atop tall towers and work like fans in reverse. Rather than using electricity to make wind, turbines use wind to make electricity.

Wind turns the blades and the blades spin a shaft that is connected through a set of gears to drive an electrical generator. Large-scale turbines for utilities can generate from 750 kilowatts (a kilowatt is 1,000 watts) to 1.5 megawatts (a megawatt is 1 million watts). Homes, telecommunications stations, and water pumps use single small turbines of less than 100 kilowatts as an energy source, particularly in remote areas where there is no utility service.

In wind plants or wind farms, groups of turbines are linked together to generate electricity for the utility grid. The electricity is sent through transmission and distribution lines to consumers.

Since 1980, research and testing sponsored by the DOE Wind Program has helped reduce the cost of wind energy from 80 cents (current dollars) per kilowatt hour to between 4 and 6 cents per kilowatt hour today.

One goal of the Wind Program is to further reduce the cost of utility-scale wind energy production to 3 cents per kilowatt hour at land-based, low-wind-speed sites and 5 cents per kilowatt hour for offshore (ocean) sites. A low-wind-speed site is one where the annual average wind speed measured 10 meters above the ground is 13 miles per hour. . . .

Global wind energy capacity has increased 10-fold in the last 10 years—from 3.5 gigawatts (a gigawatt is 1 billion watts) in 1994 to nearly 50 gigawatts by the end of 2004. In the United States, wind energy capacity tripled, from 1,600 megawatts in 1994 to more than 6,700 megawatts by the end of 2004—enough to serve more than 1.6 million households.

In 2005, because of a federal production tax credit renewed by Congress in 2004, the U.S. wind energy industry is poised for record growth. The tax credit provides a 1.9-cent per kilowatt hour credit for eligible technologies for the first 10 years of production. Some wind industry experts predict that wind technology installations for 2005 will add more than 2,000 megawatts of capacity because of the tax advantage provided by this law.

The wind industry has grown phenomenally in the past decade thanks to supporting government policies, and the work of DOE Wind Program researchers in collaboration with industry partners to develop innovative cost-reducing technologies, cultivate market growth, and identify new wind energy applications.

FAST FACT

According to the US Department of Energy, the nation has enough wind power capacity to serve more than 9 million homes and avoid annual emissions of 62 million tons of carbon dioxide.

Developing Cost-Reducing Technologies

Work conducted under DOE Wind Program projects from 1994 to 2004 produced innovative designs, larger turbines, and efficiencies that have led to dramatic cost reductions. Although this drop in cost is impressive, electricity produced by wind energy is not yet fully competitive with that produced by fossil fuels. Researchers believe that further technology improvements will be needed to reduce the cost of electricity from wind another 30 percent for it to become fully competitive with conventional fuel-consuming electricity generation technologies.

Cultivating Market Growth

To cultivate market growth by increasing acceptance of wind technology around the country, DOE's Wind Powering America (WPA)

The American Wind Energy Association states that for every 10,000 birds killed by human activities, less than one death is caused by a wind turbine.

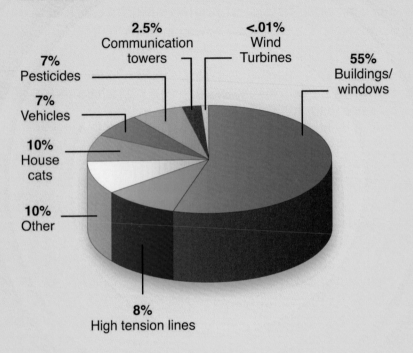

2.5% Communication towers

<.01% Wind Turbines

7% Pesticides

55% Buildings/ windows

7% Vehicles

10% House cats

10% Other

8% High tension lines

Taken from: American Wind Energy Association.

team works with industry partners to provide state support, develop utility partnerships, conduct outreach, and develop innovative market mechanisms to support the use of large- and small-scale wind systems. . . .

Through such efforts, the WPA seeks to increase the use of wind energy in the United States with the goal that at least 30 states have 100 megawatts of wind capacity by 2010.

New Wind Energy Applications

Decades of work conducted through public-private partnerships have moved wind energy from yesterday's dream to today's reality. . . .

Offshore and Deep-Water Development

Offshore wind turbines, now in the early stages of development, are more expensive and harder to install and maintain than turbines on land. Offshore turbines must be designed to survive the offshore wind and wave loading of severe storms, and protected from the corrosive marine environment.

Some advantages of offshore installation are that turbines can be made bigger than those onshore to produce more power per turbine, and the ocean location provides greatly increased wind speeds and less turbulence. Offshore installations also reduce land-use and could ease aesthetic concerns, if the turbines are located far from shore and out of sight.

Recent studies show that there are significant offshore wind resources in regions of the United States near major urban areas in the mid-Atlantic and northeast. In Europe, offshore wind turbines produce about 600 megawatts, but no turbines have yet been installed in waters deeper than 20 meters.

For offshore turbines in shallow water (less than 30 meters), European turbine manufacturers have adopted conventional land-based turbine designs and placed them on concrete bases or steel monopiles driven into the seabed. An offshore substation collects the energy and boosts the voltage, and then a buried undersea cable carries the power to shore, where another substation provides a further voltage increase for transmission to utilities for distribution to customers.

A large amount of potential U.S. offshore wind resources are in waters deeper than the current technology limit of about 30 meters, as developed in Europe for the Baltic Sea. Monopile foundations driven into the seabed are less suitable for the deeper waters off U.S. coasts. To produce cost-effective wind energy in deep water, floating platform technologies developed by the oil and gas industries need to be adapted and scaled for application to wind energy and new lower-cost anchoring methods developed. The ultimate vision for this new offshore wind technology would be to build the turbines and the supporting platform in a shore-based dry dock with local labor, tow the floating turbine to its place on the sea, drop anchor, and plug in to the power cable to shore. . . .

The world's largest wind farm was constructed off the coast of Denmark in 2009. Denmark gets approximately 15 percent of its total energy needs from wind power.

Wind and Water

The Wind Program is investigating how wind and water can work together to provide a more stable supply of electricity and fresh water. The scarcity of fresh water is a growing global problem. According to the United Nations, the world's burgeoning population will need billions more cubic meters of fresh water per day by 2025. The current global desalination capacity is an estimated 28 million cubic meters per day.

An important solution to water scarcity is desalination of abundant ocean salt water, but desalination is a highly energy-intensive technology and is not cost effective in most areas. Among all the desalination process technologies, reverse osmosis has the highest electrical energy efficiency, at 3-8 kilowatt hours per cubic meter of water.

Reverse osmosis is a method of producing pure water by forcing salt water through a semipermeable membrane (which allows some molecules through but not others) that salts cannot pass through.

Even with the high efficiency of reverse osmosis, energy accounts for about 40 percent of the total desalinated water cost. From a cost and environmental point of view, inexpensive and clean alternative power sources are needed for a low-cost desalination solution.

Wind power is the most promising and least expensive renewable power source but, because of its variable nature—because wind doesn't always blow—researchers must determine the effects it will have on desalination systems and their operation. . . .

Wind Can Be Our Future

The U.S. Department of Energy's Program to make clean and sustainable wind energy cost effective for several market applications has made significant progress in recent years and is on a steady course to further significant improvements. Sound and sustainable development of this renewable energy resource is a key element of the U.S. strategy to reduce national reliance on carbon-based fuels and reduce the production of greenhouse gas emissions.

EVALUATING THE AUTHORS' ARGUMENTS:

Compare this viewpoint with the one by Ed Hiserodt, who argues that the only viable sources for large-scale energy production in the future are natural gas, nuclear power, and coal. Which view do you agree with most? Discuss your answer using evidence from each viewpoint.

Solar Power Is a Viable Alternative Energy Source

Jeremy Leggett

"The case for solar power is simple: capture only a tiny fraction of 1 per cent of the sun's energy and we could provide much more power than the world needs."

In the following viewpoint Jeremy Leggett argues that solar power is a viable alternative energy source. Leggett argues that Earth receives much more sunlight than is required to power the whole world. If that sunlight can be efficiently harnessed—which Leggett believes it can—every country in the world can benefit. Leggett acknowledges that switching to a solar-based energy economy will be expensive but argues that the rising cost of oil makes the added expense negligible in just a few years. Making the switch is worth it: Leggett points out that solar energy is clean, renewable, and would help the world's nations enjoy a more stable and less volatile existence. For all these reasons he urges nations around the world to invest in solar technology. Leggett is a geologist and the author of *The Solar Century* (2009) and *The Empty Tank: Oil, Gas, Hot Air, and the Coming Global Financial Catastrophe* (2005).

Jeremy Leggett, "Hello Sunshine," *Prospect Magazine*, August 27, 2009. Reprinted by permission.

AS YOU READ, CONSIDER THE FOLLOWING QUESTIONS:
1. What is the Desertec plan, as described by Leggett?
2. How much of the United States' needed electricity does Leggett say could be generated by just two hundred square kilometers of solar plants?
3. In how many years does the author say the United States would see a payback on its investment in solar power, given the rising price of oil?

The world needs about 13 terawatts of power every year, a figure that will rise to about 20 terawatts by 2020. The amount of sunlight falling on the planet at any one time is around 120,000 terawatts—more than 9,000 times what we need. The case for solar power is simple: capture only a tiny fraction of 1 per cent of the sun's energy and we could provide much more power than the world needs. Add a little energy efficiency and some renewable energy, and we can do it with ease.

Power Europe with Just 0.3 Percent of Desert Sunlight

Where to start? The obvious places are the parts of the planet that get the most sun already: deserts. With the right technology electricity could be transported from hot countries on gigantic international energy grids for use in cloudy countries. Most of us associate solar energy with solar panels, otherwise known as solar photovoltaics (PV). But another widely applied technology uses the sun to heat up fluids—in power stations, for example, to heat water to drive turbines—an approach known as concentrating solar power technology (CSP).

To supply all the electricity Europe needs would, in principle, mean capturing just 0.3 per cent of the light falling on the Saharan and middle eastern deserts, in an area smaller than Wales. Knowing this, a group of German industrial giants in July [2009] launched what may be the world's largest solar project: a long-planned partnership of 20 finance and energy companies to build new African CSP plants, led by Munich Re, and including E.ON, Siemens and Deutsche Bank. Picture row upon row of curved mirrors with semicircular profiles, in glittering fields, scattered across the countries of northern Africa. The

In 2009 a German industrial consortium announced the Desertec plan to build the world's largest concentrated solar power facilities, like the one shown, in North Africa. It will produce 15 percent of Europe's energy needs.

plan—dubbed Desertec—hopes ultimately to provide 15 per cent of Europe's electricity needs, and aims to figure out how this can be done within the next two to three years.

Such a giant project will not be easy. The estimated cost, somewhere near a cool €555bn [billion euros] ($788bn), might seem a lot. But given that the International Energy Agency calculates that we need to spend $45 trillion on new energy systems over the next 40 years, it could represent good value. The real problem is getting the power back to Europe. The plan has backing from the European commission, whose Institute for Energy envisages a network of African CSP plants linked to Europe by a new high-voltage transmission grid, which could also capture energy from wind power from across Europe and western North Africa. Building such a grid, using overland lines and submarine cables, is technically challenging, but far from impossible.

Whole Nations Could Run on Sun Alone

With the grid in place, Desertec-type schemes could in the future even run entire nations on solar energy alone. One 2006 study calculated that all US electricity could be provided by covering 200 sq km with new solar plants. The plants would have to be in America's sunny southwest, and power transmitted to the cooler north. This isn't as inefficient as it sounds: electricity use is high in summer in the northeast, when air conditioning is needed, and this is just the time when solar generation is at its maximum in the US southwest.

All this would cost the US between $4.5-6 trillion in capital investment, at today's prices. But at an oil price of $100 a barrel—to take a conservative estimate for the coming decades—the 13m barrels a day that the US imported in 2006 would cost almost half a trillion dollars a year. The "payback" on avoided oil imports would therefore come in just nine to 12 years, with coal and gas savings factored in.

In December 2007 three respected researchers published a plan in the magazine *Scientific American* that, by 2050, would see solar photovoltaic farms produce 3,000 giga-watts in the American southwest, plus some large Desertec-type solar thermal plants. Excess energy produced during the daytime would be stored as compressed air in underground caverns, which could then be used at night to drive turbines. With a new grid, such a plan could see solar electricity provide 69 per cent of the US's electricity by 2050. And plans for just such a grid, known

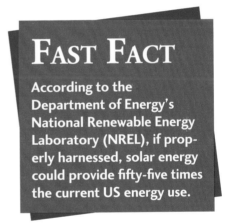

FAST FACT

According to the Department of Energy's National Renewable Energy Laboratory (NREL), if properly harnessed, solar energy could provide fifty-five times the current US energy use.

as the $10bn "green power express" (built to carry wind power in the first instance) were given funding during President [Barack] Obama's stimulus package in February [2009].

Switching to Solar Is Possible

When one understands the enormous scope that smart grids and energy storage have, it becomes easier to believe that a fundamental switch to solar and other renewables is possible. Yet challenges still

Solar Energy Is Abundant

Most regions of the United States receive large amounts of sunlight that proponents of solar energy say can be successfully harnessed and turned into power.

kWh/m2/Day (kilowatt-hour per square meter per day)

<8.3   7.0   6.0   5.0   4.0   3.0   2.0   >1.3

Taken from: National Renewable Energy Laboratory and US Department of Energy, October 2008.

remain. US domestic solar plans, in particular, have an advantage over rival European ventures like Desertec: political stability. A world that relied on foreign solar grids would need to be much less volatile than today, and multilateral institutions more robust. That could happen, of course, especially if the US and China synchronise their interests. But history suggests caution: rich nations will surely balk at replacing dependency on oil in sunny lands with solar from those same sunny lands.

Thankfully, photovoltaics works in cloudy countries too, generating energy right where it is needed, as any survey of rooftops in Germany now shows. We would be unwise to put all our eggs in one giant solar basket, as some of the mere enthusiastic engineers appear prepared to. The good news is that we don't have to.

EVALUATING THE AUTHOR'S ARGUMENTS:

In this viewpoint, Jeremy Leggett uses facts, statistics, examples, and reasoning to make his argument that solar power is a viable energy source. He does not, however, use any quotations to support his point. If you were to rewrite this article and insert quotations, what authorities might you quote from? Where would you place them, and why?

Neither Wind nor Solar Power Are Viable Alternative Energy Sources

Ed Hiserodt

"Though alternative energy options like solar and wind power continue to get favorable press, their large drawbacks limit their practicality."

In the following viewpoint, Ed Hiserodt discusses the practicality of energy production from wind, solar, hydropower, natural gas, coal, and nuclear sources. Hiserodt describes wind and solar power as inefficient, subject to disruptions from the environment, and requiring large tracts of land for their energy-generating machinery. According to Hiserodt, hydropower has peaked and some dams will likely be dismantled. Natural gas has been used increasingly in the last decade, and its prices have increased also. Currently, either new natural gas sources will need to be found or gas for power generation will take gas away from other consumers of gas. Coal and nuclear sources are the most viable for

Ed Hiserodt, "Another Look at Nuclear Energy: Nuclear Energy Is on the Go, Helping Countries in Europe and Other Parts of the World Solve Their Energy Woes Economically and Safely. Will America Get Back on Board?," *The New American*, April 30, 2007. All rights reserved. Reproduced by permission.

power generation, but, according to Hiserodt, nuclear is safer, more economical, and requires a smaller quantity of fuel. Ed Hiserodt is an aerospace engineer, an expert in power generation technology, and the president of Controls & Power, Inc. since 1983.

AS YOU READ, CONSIDER THE FOLLOWING QUESTIONS:

1. According to the author and *Wired* magazine, how much land will Stirling Energy Systems need to build its 500-megawatt solar power plant in Southern California?
2. How much of America's electricity does coal produce, and why is it a good option for future power generation, according to the author?
3. According to the author, for similar one-thousand-megawatt power plants, how many ninety-ton train cars of coal does it take to power a coal-powered electrical plant for a day? How many truckloads of uranium to fuel a nuclear power plant for a year?

The site of what is arguably the world's leading research program in nuclear energy lies just a short drive from the city of Marseille through the picturesque and romantic countryside of southern France. At Cadarache, the Commissariat a l'energie atomique, the French atomic energy agency, operates a complex of research facilities that is soon to be the home of ITER, the International Thermonuclear Experimental Reactor, which will be the most advanced and powerful Tokamak fusion reactor ever built. This reactor is being funded by the EU, China, Russia, the United States, and others.

Though famous as the future site of ITER, Cadarache is also soon to be home to one of the world's most advanced fission reactors. On March 21, [2007,] engineers began building the advanced Jules Horowitz Reactor (JHR), to be used to test and evaluate advanced technologies. It is expected to operate for 50 years.

Work on the JHR is just one more sign of French dominance in the nuclear energy industry. In the United States, where nuclear energy technology was invented, only 19.4 percent of electricity is supplied by nuclear power plants. In France, by comparison, 78.5 percent of electricity is generated by nuclear power. The situation is much the same in other European nations. Lithuania, Slovakia, and Belgium all generate more than half of their electricity using nuclear power. Other nations producing more than 40 percent of their electricity using nuclear power include Ukraine, Sweden, Bulgaria, Armenia, and Slovenia. Meanwhile, a growing list of nations, including Russia, China, South Korea, Taiwan, and India, have nuclear power plants under construction.

Notably absent from that list is the United States. Hamstrung by irrational fears, miles of red tape, and onerous bureaucratic regulatory obstacles, no new domestic nuclear power plants have been ordered and built in America for over 30 years, even though the United States has the largest per capita demand for energy. The resulting lack of new nuclear capacity in the face of rising energy demand brings on short-term and long-term consequences. Short-term problems caused by a lack of electrical capacity include rolling blackouts, disruptions in daily life like stopped elevators, non-working traffic signals, loss of refrigerated products, etc. Among long-term problems, we face rising electrical costs, termination of marginal industries, and no industrial expansion, among many, many others. Those consequences are entirely unnecessary because nuclear power provides an economical and safe method of producing abundant electricity.

Energy Options

Though alternative energy options like solar and wind power continue to get favorable press, their large drawbacks limit their practicality. Both are inefficient and subject to environmental disruptions. Cloudy days substantially reduce output from solar "farms," and wind power is subject to the vagaries of the weather. Moreover, vast swaths of territory must be covered with solar panels and parabolic mirrors or windmills in order to generate large amounts of power. In southern California, for instance, Stirling Energy Systems is building a 500-megawatt solar installation that, according to *Wired* magazine, is expected to cover 4,500 acres of land with 20,000 large, dish-shaped mirrors.

The only really viable alternatives for future large-scale generation lie with the four technologies that already provide the bulk of the nation's power: hydropower, natural gas, nuclear, and coal. Of those, hydropower, providing 6.5 percent of U.S. electricity, has already peaked, and is likely to undergo a slow decline in usage in the future as smaller dams are dismantled to restore the natural course of some rivers. That leaves natural gas, coal, and nuclear power as options.

While natural gas has been highly touted as an energy source because it is considered a relatively clean fuel and because gas plants are relatively inexpensive to build, gas is an unlikely candidate for future large-scale power generation. During the 1990s, construction of gas-fired plants, which now provide 18.7 percent of U.S. electricity, increased because of relatively low fuel prices. In recent years, however, natural gas prices have increased substantially.

Moreover, domestic gas supplies are likely to be insufficient to support long term expansions in gas-fired generation, even if domestic gas production increases. A recent study by Canada's National Energy Board concluded: "Production increases alone are not sufficient to meet the projected future requirements for natural gas demand, including power generation. Consequently, any increases in demand for gas-fired generation would necessitate a reduction in gas consumption by other consumers and the development of further sources of gas supply." Those other sources of supply would be foreign, most likely Russian, sources. Already Europe is dependent on Russian gas and in each of the last two years faced supply disruptions at the hands of the Kremlin.

Gas isn't going away any time soon, but it is clearly not the best solution to the nation's long term need for energy. Coal, which already provides almost 50 percent of U.S. electricity, is a better option because the United States holds the world's largest reserves of the fuel. But even coal—though it will remain a useful fuel far into the future—fares poorly in comparison to nuclear energy.

Not far from my home, one nearby road crosses a rail line that leads to a coal-fired plant. As with most rail lines, trains here seem to be synchronized to run at times of maximum inconvenience. Most of these trains on this line are known as "unit trains," consisting of an engine or two plus one hundred 90-ton coal cars. It takes all the coal hauled by one of these trains to fuel a typical 1,000-megawatt generating plant for a single day! And even though the coal itself is cheap at the mouth of the mine, transportation costs mount substantially by the time the unit train reaches a power plant. According to the University of Wyoming, "Coal at the mine mouth is about $5 per ton. By the time it gets to Illinois, the cost is $30 per ton. A train load of coal is worth $50,000 when it leaves the mine. When it pulls into the power plant in Chicago it is worth $300,000! For the user, up to 80% of the cost of the coal is in the transportation." . . .

The Nuclear Option

Even though coal remains an attractive option, nuclear energy is far superior. For one thing, despite the bad press it gets, nuclear is safer. No one in the United States has died as a result of nuclear-power generation. That can't be said for coal. Historically, more than 100 lives have been lost annually at train crossings owing to coal-hauling unit trains. . . .

All things being equal, the economics of energy production also favor nuclear energy over coal. Instead of the 100 or so train cars of coal it takes to run the average coal plant each day, nuclear energy uses a comparatively tiny amount of uranium for fuel, making nuclear energy very efficient by comparison. The relatively tiny fuel requirements of nuclear power plants result in operational cost savings, and new technology developed by a private team of scientists in Australia and leased to General Electric promises to reduce costs even more. . . .

Today, most existing nuclear power plants require uranium fuel that is comprised of about 3 percent U235, with the balance being the more abundant U238 isotope. All told, it takes just six truckloads of uranium to power a typical 1,000-megawatt nuclear reactor for a year. . . .

EVALUATING THE AUTHOR'S ARGUMENTS:

What are the criteria that Ed Hiserodt uses to determine viability and practicality? Do you agree with his choice of criteria? Discuss your answer using evidence from the viewpoint.

Viewpoint

6

Nuclear Power Is a Viable Alternative Energy Source

Rob Johnston

"[Nuclear power is] the only proven safe and cost-effective way to generate large amounts of electricity that won't produce large amounts of greenhouse gas emissions."

Nuclear power is the only true alternative energy source contends Rob Johnston in the following viewpoint. He argues that nuclear power is cheaper, more efficient, and more powerful than all the other alternatives to fossil fuels, such as wind or hydropower. He dispels myths that nuclear power plants cause cancer or are a target for terrorists. Johnston also says there are hundreds of years of nuclear fuel in the earth, and so nuclear power, though technically not a renewable resource, can be relied upon for centuries to come. Johnston concludes that nuclear power alone is capable of delivering the quantities of energy the world needs at a price it can afford.

Johnston is a UK-based writer. He frequently comments on the environment, health, and science topics.

Rob Johnston, "Ten Myths About Nuclear Power," *Spiked,* January 9, 2008. Reprinted by permission.

G reens opposing nuclear power muddle every issue from terrorism to uranium supplies, in order to besmirch the only proven safe and cost-effective way to generate large amounts of electricity that won't produce large amounts of greenhouse gas emissions. One would think that greens don't want a world with abundant energy *and* a stable climate! . . .

Nuclear Fuel Supply Will Last Centuries

According to Greenpeace, uranium reserves are 'relatively limited' and last week [January 2008] the Nuclear Consultation Working Group claimed that a significant increase in nuclear generating capacity would reduce reliable supplies from 50 to 12 years.

In fact, there is 600 times more uranium in the ground than gold and there is as much uranium as tin. There has been no major new uranium exploration for 20 years, but at current consumption levels, known uranium reserves are predicted to last for 85 years. Geological estimates from the International Atomic Energy Agency (IAEA) and the Organisation for Economic Cooperation and Development (OECD) show that at least six times more uranium is extractable—enough for 500 years' supply at current demand. Modern reactors can use thorium as a fuel and convert it into uranium—and there is three times more thorium in the ground than uranium.

Uranium is the only fuel which, when burnt, generates *more* fuel. Not only existing nuclear warheads, but also the uranium and plutonium in radioactive waste can be reprocessed into new fuel, which former UK chief scientist Sir David King estimates could supply 60 per cent of Britain's electricity to 2060.

In short, there is more than enough uranium, thorium and plutonium to supply the entire world's electricity for several hundred years. . . .

Nuclear Power Is Cheaper than Other Energy Alternatives

With all power generation technology, the cost of electricity depends upon the investment in construction (including interest on capital loans), fuel, management and operation. Like wind, solar and hydroelectric dams, the principal costs of nuclear lie in construction. Acquisition of uranium accounts for only about 10 per cent of the price of total costs, so nuclear power is not as vulnerable to fluctuations in the price of fuel as gas and oil generation.

Unlike the UK's existing stations, any new designs will be pre-approved for operational safety, modular to lower construction costs, produce 90 per cent less volume of waste and incorporate decommissioning and waste management costs.

A worst-case analysis conducted for the UK Department of Trade and Industry (now the Department of Business and Enterprise), which was accepted by Greenpeace, shows nuclear-generated electricity to be only marginally more expensive than gas (before the late-2007 hike in gas prices), and 10 to 20 times cheaper than onshore and offshore wind. With expected carbon-pricing penalties for gas and coal, nuclear power will be considerably cheaper than all the alternatives.

> **FAST FACT**
>
> According to the US Energy Information Administration, in 2009, nuclear energy provided about 9 percent of US total energy needs and about 20 percent of its total electricity needs.

Reactor Waste Is Minimal

Contrary to environmentalists' claims, Britain is not overwhelmed with radioactive waste and has no radioactive waste 'problem'.

By 2040 there will be a total of 2,000 cubic metres of the most radioactive high-level waste, which would fit in a 13 × 13 × 13 metre hole—about the size of the foundations for one small wind turbine.

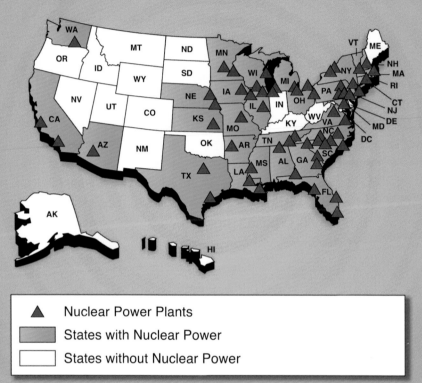

Nuclear Power in the United States

There are 104 nuclear reactors operating at sixty-five plants around the nation. Together those generate about 20 percent of the nation's electricity.

▲ Nuclear Power Plants

◼ States with Nuclear Power

☐ States without Nuclear Power

Taken from: US Energy Information Administration, September 2010.

Much of this high-level waste is actually a leftover from Britain's atomic weapons programme. All of the UK's intermediate and high-level radioactive waste for the past 50 years and the next 30 years would fit in just one Royal Albert Hall, an entertainment venue in London that holds 6,000 people (and which seems, for some reason, to have become the standard unit of measurement in debates about any kind of waste in the UK).

The largest volume of waste from the nuclear power programme is low-level waste—concrete from outbuildings, car parks, construction materials, soil from the surroundings and so on. By 2100, there will be 473,000 cubic metres of such waste from decommissioned plants—enough to fill five Albert Halls.

Production of all the electricity consumed in a four-bedroom house for 70 years leaves about one teacup of high-level waste, and new nuclear build[ing] will not make any significant contribution to existing radioactive waste levels for 20–40 years. . . .

Nuclear Reactors Do Not Cause Cancer

Childhood leukaemia rates are no higher near nuclear power plants than they are near organic farms. 'Leukaemia clusters' are geographic areas where the rates of childhood leukaemia appear to be higher than normal, but the definition is controversial because it ignores the fact that leukaemia is actually several very different (and unrelated) diseases with different causes.

The major increase in UK childhood leukaemia rates occurred before the Second World War. The very small (one per cent) annual increase seen now is probably due to better diagnosis, although it is possible that there is a viral contribution to the disease.

The Sellafield nuclear reprocessing plant is located in Cumbria, Great Britain. The author of the viewpoint argues that nuclear energy is a more practical source of energy than solar or wind power.

It is purely by chance that a leukaemia 'cluster' will occur near a nuclear installation, a national park or a rollercoaster ride. One such 'cluster' occurred in Seascale, the nearest village to the Sellafield nuclear reprocessing plant, but there are no other examples. Clusters tend to be found in isolated areas where there has been a recent influx of immigration—which hints at a virus.

Men who work on nuclear submarines or in nuclear plants are no more likely to father children with leukaemia (or any other disease) than workers in any other industry.

Reactors Do Not Lead to Weapons Proliferation

More nuclear plants (in Britain and elsewhere) would actually reduce weapons proliferation. Atomic warheads make excellent reactor fuel; decommissioned warheads (containing greatly enriched uranium or plutonium) currently provide about 15 per cent of world nuclear fuel. Increased demand for reactor fuel would divert such warheads *away* from potential terrorists. Nuclear build[ing] is closely monitored by the IAEA, which polices anti-proliferation treaties.

Better Than Renewables

If, as greens say, new nuclear power cannot come on-line in time to prevent climate change, how much less impact can wind, wave and carbon capture make?

Environmentalists claim offshore wind turbines can make a significant contribution to electricity supply. Even if that were true—which it is certainly not—the environmental impact disqualifies wind as 'sustainable'. The opening up of the North Sea continental shelf to 7,000 wind turbines is, essentially, the building of a huge industrial infrastructure across a vast swathe of ecologically sensitive seabed—as 'unsustainable' in its own way as the opening of the Arctic Wildlife Refuge [in Alaska] to oil exploration.

Wave power is still highly experimental and unproven as a method of generating electricity. Even if we allow the Severn Tidal Bore, the tidal surge that runs up and down the River Severn estuary in south-west England (and a great natural wonder of the world), to be destroyed, the cost overruns and time delays would make any problems of the nuclear industry look cheap by comparison.

Reactors Are Not a Terrorist Target

Since 11 September 2001, several studies have examined the possibility of attacks by a large aircraft on reactor containment buildings. The US Department of Energy sponsored an independent computer-modelling study of the effects of a fully fuelled Boeing 767-400 hitting the reactor containment vessel. Under none of the possible scenarios was containment breached.

Only the highly specialised US 'bunker busting' ordnance would be capable—after several direct strikes—of penetrating the amount of reinforced concrete that surrounds reactors. And besides, terrorists have already demonstrated that they prefer large, high visibility, soft targets with maximum human casualties (as in the attacks on New York, London, Madrid and Mumbai) rather than well-guarded, isolated, low-population targets.

Any new generation of nuclear reactors in the UK will be designed with even greater protection against attack than existing plants, and with 'passive' safety measures that work without human intervention or computer control.

EVALUATING THE AUTHORS' ARGUMENTS:

Rob Johnston and Kristin Shrader-Frechette, author of the following viewpoint, disagree on whether nuclear power plants are vulnerable to terrorist attack. Write one paragraph that lays out each author's argument and supporting evidence on this topic. Then, state with which author you agree. What piece of evidence swayed you?

Nuclear Power Is Not a Viable Alternative Energy Source

"In addition to being risky, nuclear power is unable to meet our current or future energy needs."

Kristin Shrader-Frechette

In the following viewpoint Kristin Shrader-Frechette argues that nuclear power is a dirty, expensive, and unsafe energy source that should not be invested in by the US government or by energy companies. She points out that although nuclear power plants do not release polluting emissions, the process by which nuclear power is generated is not only bad for the environment but requires vast amounts of fossil fuels. Furthermore, Shrader-Frechette says that nuclear power is actually more expensive than renewable sources of power, when one factors costs such as the expense of building and maintaining nuclear power plants. Finally, Shrader-Frechette argues that because nuclear power technology is prone to terrorist attack and weapons development, nuclear power puts the lives of civilians at risk. For all of these reasons she concludes that nuclear power is not a

Kristin Shrader-Frechette, "Five Myths About Nuclear Energy," *America,* June 23, 2008. Reprinted by permission.

viable alternative energy source. Shrader-Frechette teaches biological sciences and philosophy at the University of Notre Dame. She is the author of *Taking Action, Saving Lives: Our Duties to Protect Environmental and Public Health.*

AS YOU READ, CONSIDER THE FOLLOWING QUESTIONS:
1. In what way is nuclear power neither a clean nor green energy source, according to Shrader-Frechette?
2. What does the word "subsidy" mean in the context of the viewpoint?
3. Who is Hannes Alven, and how does he factor into the author's argument?

A tomic energy is among the most impractical and risky of available fuel sources. Private financiers are reluctant to invest in it, and both experts and the public have questions about the likelihood of safely storing lethal radioactive wastes for the required million years. Reactors also provide irresistible targets for terrorists seeking to inflict deep and lasting damage on the United States. The government's own data show that U.S. nuclear reactors have more than a one-in-five lifetime probability of core melt, and a nuclear accident could kill 140,000 people, contaminate an area the size of Pennsylvania, and destroy our homes and health.

The Myths That Surround Nuclear Power

In addition to being risky, nuclear power is unable to meet our current or future energy needs. Because of safety requirements and the length of time it takes to construct a nuclear-power facility, the government says that by the year 2050 atomic energy could supply, at best, 20 percent of U.S. electricity needs; yet by 2020, wind and solar panels could supply at least 32 percent of U.S. electricity, at about half the cost of nuclear power.

Nevertheless, in the last two years, the current U.S. administration [of George W. Bush] has given the bulk of taxpayer energy subsidies—a total of $20 billion—to atomic power. Why? Some officials say nuclear energy is clean, inexpensive, needed to address global climate change, unlikely to increase the risk of nuclear proliferation and safe.

On all five counts they are wrong. Renewable energy sources are cleaner, cheaper, better able to address climate change and proliferation risks, and safer. The government's own data show that wind energy now costs less than half of nuclear power; that wind can supply far more energy, more quickly, than nuclear power; and that by 2015, solar panels will be economically competitive with all other conventional energy technologies. The administration's case for nuclear power rests on at least five myths. Debunking these myths is necessary if the United States is to abandon its current dangerous energy course.

Nuclear Energy Is Not Clean or Green

The myth of clean atomic power arises partly because some sources, like a pro-nuclear energy analysis published in 2003 by several professors at the Massachusetts Institute of Technology [M.I.T.], call atomic power a "carbon-free source" of energy. On its Web site, the U.S. Department of Energy [D.O.E.], which is also a proponent of nuclear energy, calls atomic power "emissions free." At best, these claims are half-truths because they "trim the data" on emissions.

While nuclear reactors themselves do not release greenhouse gases, reactors are only part of the nine-stage nuclear fuel cycle. This cycle includes mining uranium ore, milling it to extract uranium, converting the uranium to gas, enriching it, fabricating fuel pellets, generating power, reprocessing spent fuel, storing spent fuel at the reactor and transporting the waste to a permanent storage facility. Because most of these nine stages are heavily dependent on fossil fuels, nuclear power thus generates at least 33 grams of carbon-equivalent emissions for each kilowatt-hour of electricity that is produced. (To provide uniform calculations of greenhouse emissions, the various effects of the different greenhouse gases typically are converted to carbon-equivalent emissions.) Per kilowatt-hour, atomic energy produces only one-seventh the greenhouse emissions of coal, but twice as much as wind and slightly more than solar panels.

Nuclear power is even less clean when compared with energy-efficiency measures, such as using compact-fluorescent bulbs and increasing home insulation. Whether in medicine or energy policy, preventing a problem is usually cheaper than curing or solving it, and energy efficiency is the most cost-effective way to solve the problem of reducing greenhouse gases. . . .

Nuclear Energy Is Expensive

Achieving greater energy efficiency, however, also requires ending the lopsided system of taxpayer nuclear subsidies that encourage the myth of inexpensive electricity from atomic power. Since 1949, the U.S. government has provided about $165 billion in subsidies to nuclear energy, about $5 billion to solar and wind together, and even less to energy-efficiency programs. All government efficiency programs—to encourage use of fuel-efficient cars, for example, or to provide financial assistance so that low-income citizens can insulate their homes—currently receive only a small percentage of federal energy monies.

After energy-efficiency programs, wind is the most cost-effective way both to generate electricity and to reduce greenhouse emissions. It costs about half as much as atomic power. The only nearly finished nuclear plant in the West, now being built in Finland by the French company Areva, will generate electricity costing 11 cents per kilowatt-hour. Yet the U.S. government's Lawrence Berkeley National Laboratory calculated actual costs of new wind plants, over the last seven years, at 3.4 cents per kilowatt-hour. Although some groups say nuclear energy is inexpensive, their misleading claims rely on trimming the data on cost. The 2003 M.I.T. study, for instance, included neither the costs of reprocessing nuclear material, nor the full interest costs on nuclear-facility construction capital, nor the total costs of waste storage. Once these omissions—from the entire nine-stage nuclear fuel cycle—are included, nuclear costs are about 11 cents per kilowatt-hour.

The cost-effectiveness of wind power explains why in 2006 utility companies worldwide added 10 times more wind-generated, than nuclear, electricity capacity. It also explains why small-scale sources of renewable energy, like wind and solar, received $56 billion in global private investments in 2006, while nuclear energy received nothing. It explains why wind supplies 20 percent of Denmark's electricity. It explains why, each year for the last several years, Germany, Spain and India have each, alone, added more wind capacity than all countries in the world, taken together, have added in nuclear capacity. . . .

Should the United States continue to heavily subsidize nuclear technology? Or, as the distinguished physicist Amory Lovins put it, is the nuclear industry dying of an "incurable attack of market forces"? Standard and Poor's, the credit- and investment-rating company, downgrades the rating of any utility that wants a nuclear plant. It claims that even subsidies are unlikely to make nuclear investment wise. *Forbes* magazine recently called nuclear investment "the largest managerial disaster in business history," something pursued only by the "blind" or the "biased."

Nuclear Energy Will Not Help Avert Climate Change

Government, industry and university studies, like those recently from Princeton, agree that wind turbines and solar panels already exist at an industrial scale and could supply one-third of U.S. electricity needs by 2020, and the vast majority of U.S. electricity by 2050—not just the 20 percent of electricity possible from nuclear energy by 2050. The D.O.E. says wind from only three states (Kansas, North Dakota and Texas) could supply all U.S. electricity needs, and 20 states could supply nearly triple those needs. By 2015, according to the D.O.E., solar panels will be competitive with all conventional energy technologies and will cost 5 to 10 cents per kilowatt hour. Shell Oil and other fossil-fuel companies agree. They are investing heavily in wind and solar.

FAST FACT

According to the World Nuclear Association, the United States operates 104 nuclear power reactors in thirty-one states.

From an economic perspective, atomic power is inefficient at addressing climate change because dollars used for more expensive, higher-emissions nuclear energy cannot be used for cheaper, lower-emissions renewable energy. Atomic power is also not sustainable. Because of dwindling uranium supplies, by the year 2050 reactors would be forced to use low-grade uranium ore whose greenhouse emissions would roughly equal those of natural gas. Besides, because the United States imports nearly all its uranium, pursuing nuclear power continues the dangerous pattern of dependency on foreign sources to meet domestic energy needs.

Nuclear Energy Will Increase Weapons Proliferation and Terrorism

Pursuing nuclear power also perpetuates the myth that increasing atomic energy, and thus increasing uranium enrichment and spent-fuel reprocessing, will increase neither terrorism nor proliferation of

Nuclear waste containers are loaded on a truck in Germany. Storage of nuclear waste is controversial because most people do not want waste facilities near their homes.

nuclear weapons. This myth has been rejected by both the International Atomic Energy Agency and the U.S. Office of Technology Assessment. More nuclear plants means more weapons materials, which means more targets, which means a higher risk of terrorism and proliferation. The government admits that [global terror network] Al Qaeda already has targeted U.S. reactors, none of which can withstand attack by a large airplane. Such an attack, warns the U.S. National Academy of Sciences, could cause fatalities as far away as 500 miles and destruction 10 times worse than that caused by the nuclear accident at Chernobyl in 1986.

Nuclear energy actually increases the risks of weapons proliferation because the same technology used for civilian atomic power can be used for weapons, as the cases of India, Iran, Iraq, North Korea and Pakistan illustrate. As the Swedish Nobel Prize winner Hannes Alven put it, "The military atom and the civilian atom are Siamese twins." Yet if the world stopped building nuclear-power plants, bomb ingredients would be harder to acquire, more conspicuous and more costly politically, if nations were caught trying to obtain them. Their motives for seeking nuclear materials would be unmasked as military, not civilian. . . .

Renewables Are More Viable than Nuclear Power

Despite the problems with atomic power, society needs around-the-clock electricity. Can we rely on intermittent wind until solar power is cost-effective in 2015? Even the Department of Energy says yes. Wind now can supply up to 20 percent of electricity, using the current electricity grid as backup, just as nuclear plants do when they are shut down for refueling, maintenance and leaks. Wind can supply up to 100 percent of electricity needs by using "distributed" turbines spread over a wide geographic region—because the wind always blows somewhere, especially offshore.

Many renewable energy sources are safe and inexpensive, and they inflict almost no damage on people or the environment. Why is the current U.S. administration instead giving virtually all of its support to a riskier, more costly nuclear alternative?

EVALUATING THE AUTHORS' ARGUMENTS:

Kristin Shrader-Frechette disagrees with Rob Johnston, author of the preceding viewpoint, on whether nuclear power is too expensive to be a viable replacement for fossil fuels. Write one paragraph that lays out each author's argument and supporting evidence on this topic. Then, state with which author you agree. What piece of evidence swayed you?

Chapter 3

What Fuel Should Power the Cars of the Future?

An electric car recharges with a curbside recharger. Controversy abounds on just what is the best fuel alternative for cars—hydrogen fuel cells, biofuels, or electricity.

Hydrogen Fuel Cells Could Power the Vehicles of the Future

Michael V. Copeland

"Hydrogen fuel cells are twice as efficient as gasoline engines, and even hydrogen produced by natural gas translates to 10% to 40% fewer emissions than gas-hybrid cars."

Hydrogen fuel cells are increasingly being considered a viable way to power the vehicles of the future, explains Michael V. Copeland in the following viewpoint. He discusses how automakers and consumers are showing heightened interest in hydrogen-powered vehicles because they are environmentally friendly and work well. Unlike older models, newer hydrogen cars are capable of driving hundreds of miles at a time and use small, practical-sized fuel cells. In addition, their only emission is water. Copeland admits significant obstacles must be overcome before hydrogen fuel cells can become commonplace technology. But he reports that a growing number of car companies, governments, and consumers are becoming interested in the technology, making it likely that hydrogen fuel cells will have a place in the fleet of the future. Michael V. Copeland is a technology writer whose work has appeared in *Fortune*, where this article was originally published.

Michael V. Copeland, "The Hydrogen Car Fights Back," *Fortune*, October 13, 2009. Reprinted by permission.

AS YOU READ, CONSIDER THE FOLLOWING QUESTIONS:
1. Who is Joan Ogden, and how does she factor into the author's argument?
2. What is the Clarity, as described by the author?
3. About how much will it cost to transition to hydrogen vehicles, as reported by the author?

The "Valley of Death," in auto-industry-speak, is a metaphorical desert where emerging technologies reside while car executives figure out which of the experiments ought to make their way into actual cars.

Every automotive leap forward has done time in the valley: turbochargers, fuel injectors, even gasoline-electric hybrids like Toyota's Prius. Hydrogen-fueled vehicles, the alternative-energy flavor of the month back in 2003, are the ones languishing today, along with hovercraft and other assorted concept cars.

But perhaps not for much longer. A number of auto manufacturers are renewing their push for hydrogen, and now it is looking as though hydrogen cars will make it out of vehicular Death Valley.

Last month Daimler (DAI), the German government, and several industrial companies announced a plan to build 1,000 hydrogen-fueling stations across Germany. Days later, Daimler CEO Dieter Zetsche showed off Mercedes-Benz's latest hydrogen-fueled effort, the F-Cell hatchback. Toyota (TM) this summer announced it would put hydrogen fuel-cell cars into production in 2015.

Honda (HMC), GM, and Hyundai all have hydrogen fuel-cell programs running, and Honda actually has put vehicles—heavily subsidized by the car maker, to be sure—in the hands of some real customers, as opposed to its own engineers. (GM today is focusing most of its, um, energy on the plug-in hybrid Chevy Volt, but the company says it still expects to have its fuel-cell technology ready for commercialization in 2015.)

"The automakers are making huge progress," says Joan Ogden, director of the Sustainable Transportation Energy Pathways Program at the University of California at Davis. "The popular notion that hydrogen is 20 years away—and always will be—is totally off-base."

Best in Show?

Actually, hydrogen cars don't have a perception problem—most people simply don't think about them at all.

It doesn't help hydrogen's cause that President Obama tried to cut funding (which Congress largely restored) for hydrogen auto research, a

Hydrogen Fueling Stations in the United States

As of 2010, there were more than seventy-five hydrogen fueling stations in the United States, and more than twenty are slated for construction.

State	Existing	Planned
Arizona	Phoenix, Yucca	
California	Arcata, Auburn, Burbank (2), Chino, Chula Vista, Culver City, Davis, Diamond Bar, Irvine (2), Lake Forest, Long Beach, Los Angeles (3), Oakland, Oceanside, Ontario, Oxnard, Port Hueneme, Riverside, San Jose, Santa Ana, Santa Monica, Thousand Palms (2), Torrance (4), West Los Angeles, West Sacramento	Emeryville, Fountain Valley, Harbor City, Los Angeles, Newport Beach, San Francisco, South Lake Tahoe, Torrance, Westwood
Colorado	Golden	Fort Collins
Connecticut	South Windsor, Wallingford	Wallingford
Delaware	Newark	Claymont
DC	Washington	
Florida	Orlando, Oviedo	Miami, Orlando
Georgia		Savannah
Hawaii	Honolulu	
Illinois	Des Plaines	Chicago
Indiana	Crane	
Maine	Scarborough	
Massachusetts	Billerica	Braintree
Michigan	Ann Arbor, Dearborn, Detroit, Milford, Romeo, Selfridge, Southfield, Taylor	
Nevada	Las Vegas (2)	
New Jersey	East Amwell	
New Mexico	Taos	
New York	Ardsley, Bronx, Hempstead, Honepye Falls, Jamaica, Latham, Rochester, White Plains	
North Carolina		Charlotte
North Dakota	Minot	
Ohio	Columbus	Brookville
Pennsylvania	Allentown, University Park	
South Carolina	Aiken, Columbia	
Texas	Austin	
Vermont	Burlington	
Virginia	Ft. Belvoir	Richmond
West Virginia	Charleston	

Taken from: FuelCells.org, 2010.

A hydrogen fuel car is refilled at a hydrogen filling station. Hydrogen fuel cells are increasingly considered a viable way to power vehicles.

pet project of the previous administration. Obama's team favors battery-powered models. So do officials in other countries, environmentalists, technology pundits, and a few prominent auto manufacturers—most notably Nissan and Ford (F, Fortune 500). And hybrids are already widely available, so consumers can judge the technology for themselves.

But hydrogen cars are about to get a bit more visible. Honda, probably hydrogen's biggest proponent, is leasing its hydrogen-powered model, the Clarity, to nine drivers in Southern California. Honda plans to make 200 Claritys available in Japan and the U.S. in the next two years.

The Clarity shows some of the improvements in hydrogen technology: It has a range of 240 miles, up from 190 miles on the 2002 model, and boasts a microwave oven-size fuel cell (that's where hydrogen combines with oxygen to produce electricity that then drives the car's motor). In the early days fuel cells were the size of filing cabinets, not exactly practical for tooling around town.

GM, Hyundai, and BMW are focused on testing hydrogen in Southern California, and soon Mercedes will be too. That is partly because many of the automakers have their R&D and design centers in the Los Angeles area, where four easily accessible hydrogen filling stations are already strategically placed close to key residential and commercial centers.

The U.S. has 64 hydrogen stations, which are owned and operated by energy companies, universities, local governments, transit agencies, and utility companies. Another 38 are well into the planning and development phase.

Actress Jamie Lee Curtis lobbied to get one of the first available models for environmental reasons, but she has become a fan of driving the Clarity. "I am not the most light-footed driver, and this thing is like a rocket ship," says Curtis, who leases the car for $600 a month. When asked what she will do when her three-year lease expires, Curtis pauses a moment. "Cry," she says. "Sob uncontrollably, and beg them to extend the lease."

Her husband, actor and film director Christopher Guest ("Best in Show"), is equally enamored, if less dramatic. "This is the best car I have ever had in terms of environmental benefits and range," says Guest, who drove a Prius until the Clarity was delivered. "I have thought about what would replace it, and there is nothing on the market that would remotely compare."

Environmental Equilibrium

Despite Curtis's and Guest's endorsements (they are not paid by Honda, but the automaker subsidizes the cost of the vehicle, estimated at $300,000, as it does for every driver), hydrogen-fueled cars still face considerable hurdles to mainstream acceptance.

One major issue will be figuring out how to produce hydrogen in a clean enough way that it doesn't offset the environmental benefits of driving a car whose only tailpipe emission is water.

While you can produce hydrogen via electrolysis from any source of electricity, including renewable-energy sources like solar and wind, it's most often extracted from natural gas combined with steam, which forms hydrogen and carbon monoxide. The carbon monoxide is then separated from the hydrogen.

Hydrogen naysayers will immediately remind you that natural gas is yet another habit-forming fossil fuel. (The "reformation" of fossil fuels produces greenhouse gases.)

Proponents of hydrogen vehicles point out that hydrogen fuel cells are twice as efficient as gasoline engines, and even hydrogen produced by natural gas translates to 10% to 40% fewer emissions than gas-hybrid cars, according to studies from MIT and the Argonne National Lab. "The societal benefits of a fuel-cell fleet would be better than a plug-in hybrid and about equal with all-battery electric," says UC Davis's Ogden.

An even bigger problem for hydrogen: lack of infrastructure. To get more hydrogen filling stations installed, car makers will need financial support from corporate partners and local and national governments. Eight of the world's major automakers—including Renault, Nissan, and Ford, which have scrapped fuel cells in favor of battery-electric options—recently called on fuel companies and regulators to expedite the rollout of hydrogen filling stations globally.

None of this will be cheap: A study commissioned by the National Academy of Science concluded that the U.S. would need to spend $3 billion to $4 billion a year for 15 years to subsidize the cost of the cars and get a national infrastructure in place to make the transition to hydrogen. Not a pittance, but to put that number in perspective, corn-based ethanol receives about the same amount in annual subsidies.

Despite the challenges hydrogen cars face, car executives maintain that they will be part of the mix of technologies shuttling us around in the future, along with gasoline-hybrid and battery-electric models. "It's not a choice of one or the other from a technology perspective," says Charlie Freese, a GM executive who heads up that company's fuel-cell efforts. "Meeting our transportation needs and environmental goals is going to require a variety of solutions." For hydrogen cars, that kind of thinking sounds like a path out of the valley of death and onto the open road.

EVALUATING THE AUTHORS' ARGUMENTS:

The author of this viewpoint, Michael V. Copeland, is a seasoned technology writer for commercial magazines. The author of the following viewpoint, Robert Zubrin, is an aerospace engineer. Does knowing the background of these authors influence your opinion of their arguments? Are you more inclined to agree with one over the other? If so, why?

Hydrogen Fuel Cells Are Unlikely to Power the Vehicles of the Future

Robert Zubrin

"America needs to abandon, once and for all, the false promise of the hydrogen age."

Robert Zubrin is an aerospace engineer and the president of Pioneer Astronautics, a research and development firm. In the following viewpoint he argues that hydrogen is not a practical energy source with which to fuel future vehicles. Zubrin says hydrogen is outrageously expensive, dangerously unstable, and highly flammable. A fuel tank that would safely hold enough of it to power a vehicle would weigh as much as a car in itself, he warns. Zubrin says the logical alternative to using pure hydrogen is to use a hydrogen fuel cell, which combines hydrogen with oxygen while a car is operating. But Zubrin says that these fuel cells have found little practical application outside of the space program, for which they were originally designed. Finally, Zubrin

Robert Zubrin, "The Hydrogen Hoax," *New Atlantis 15*, Winter 2007, pp. 9–20. Reprinted by permission.

says that hydrogen is not a truly green fuel—although it only emits water, it takes fossil fuel energy to produce it. For all of these reasons Zubrin concludes that the United States should abandon hydrogen as a possible fuel source for vehicles.

AS YOU READ, CONSIDER THE FOLLOWING QUESTIONS:
1. What does Zubrin mean when he says that hydrogen is a carrier, rather than a source, of energy?
2. How much does Zubrin say a kilogram of hydrogen costs? How does this compare with a gallon of gasoline?
3. How much would a hydrogen tank weigh, according to Zubrin?

Nearly everyone in American politics believes we face an energy crisis, and nearly everyone believes we need a technological solution that will make America "energy independent." Americans are, as President [George W.] Bush put it in his 2006 State of the Union address, "addicted to oil," and in this case our addiction is enriching and empowering those who seek to destroy us. . . . To cure this self-destructive addiction, the Bush administration has placed a major bet on the so-called "hydrogen economy," both in policy and in rhetoric. . . .

It certainly sounds great. Hydrogen, after all, is "the most common element in the universe," as [former] Energy Secretary [Spencer] Abraham pointed out. Since it is so plentiful, surely President Bush must be right when he promises it will be cheap. And when you use it, the waste product will be nothing but water—"environmental pollution will no longer be a concern." Hydrogen will be abundant, cheap, and clean. Why settle for anything less?

Unfortunately, it's all pure bunk. To get serious about energy policy, America needs to abandon, once and for all, the false promise of the hydrogen age. . . .

Hydrogen Is Not an Energy "Source"
Hydrogen is only a source of energy if it can be taken in its pure form and reacted with another chemical, such as oxygen. But all the hydrogen on Earth, except that in hydrocarbons, has already been

oxidized, so none of it is available as fuel. If you want to get plentiful unbound hydrogen, the closest place it can be found is on the surface of the Sun; mining this hydrogen supply would be quite a trick. . . .

So if we put aside the spectacularly improbable prospect of fueling our planet with extraterrestrial hydrogen imports, the only way to get free hydrogen on Earth is to make it. The trouble is that making hydrogen requires more energy than the hydrogen so produced can provide. Hydrogen, therefore, is not a *source* of energy. It simply is *a carrier* of energy. And it is, as we shall see, an extremely poor one.

Hydrogen Is Too Expensive to Use as a Fuel

The spokesmen for the hydrogen hoax claim that hydrogen will be manufactured from water via electrolysis. It is certainly possible to make hydrogen this way, but it is very expensive—so much so, that only four percent of all hydrogen currently produced in the United States is produced in this manner. The rest is made by breaking down hydrocarbons, through processes like pyrolysis of natural gas or steam reforming of coal.

Neither type of hydrogen is even remotely economical as fuel. The wholesale cost of commercial grade liquid hydrogen (made the cheap way, from hydrocarbons) shipped to large customers in the United States is about $6 per kilogram. High purity hydrogen made from electrolysis for scientific applications costs considerably more. Dispensed in compressed gas cylinders to retail customers, the current price of commercial grade hydrogen is about $100 per kilogram. For comparison, a kilogram of hydrogen contains about the same amount of energy as a gallon of gasoline [which costs around $3 per gallon]. This means that even if hydrogen cars were available and hydrogen stations existed to fuel them, no one with the power to choose otherwise would ever buy such vehicles. This fact alone makes the hydrogen economy a non-starter in a free society.

And even if you are among those willing to sacrifice freedom and economic rationality for the sake of the environment, and therefore prefer hydrogen for its advertised benefit of reduced carbon dioxide emissions, think again. Because hydrogen is actually made by reforming hydrocarbons, its use as fuel would not reduce greenhouse gas emissions at all. In fact, it would greatly increase them. . . .

An Impractical and Unsafe Fuel

In order for hydrogen to be used as fuel in a car, it has to be stored in the car. As at the station, this could be done either in the form of cryogenic liquid hydrogen or as highly compressed gas. In either case, we come up against serious problems caused by the low density of hydrogen. For example, if liquid hydrogen is the form employed, then storing 20 kilograms onboard (equivalent in energy content to 20 gallons of gasoline) would require an insulated cryogenic fuel tank with a volume of some 280 liters (70 gallons). This cryogenic hydrogen would always be boiling away, which would create concerns for those who have to leave their cars parked for any length of time, and which would also turn the atmospheres in underground or otherwise enclosed parking garages into explosive fuel-air mixtures. Public parking garages containing such cars could be expected to explode regularly, since hydrogen is flammable over concentrations in air ranging from 4 to 76 percent, and the minimum energy required for its ignition is about one-twentieth that required for gasoline or natural gas.

Compressed hydrogen is just as unworkable as liquid hydrogen. If 5,000 psi [pounds per square inch] compressed hydrogen were employed, the tank would need to be 650 liters (162 gallons), or eight times the size of a gasoline tank containing equal energy. Because it would have to hold high pressure, this huge tank could not be shaped in an irregular form to fit into the vehicle's empty space in some convenient way. Instead it would have to be a simple shape like a sphere or a domed cylinder, which would make its spatial demands much more difficult to accommodate, and significantly reduce the usable vehicle space within a car of a given size. If made of (usually) crash-safe steel, such a hydrogen tank would weigh 1,300 kilograms (2,860 pounds)— about as much as an entire small car! Lugging this extra weight around would drastically increase the fuel consumption of the vehicle, perhaps doubling it. If, instead of steel, a lightweight carbon fiber overwrapped tank were employed to avoid this penalty, the car would become a deadly explosive firebomb in the event of a crash.

Hydrogen Fuel Cells Offer Little to Vehicles

While hydrogen gas can be used as a fuel in internal combustion engines, there is no advantage in doing so. In fact, hydrogen reduces the efficiency of such engines by 20 percent compared to what they can achieve using gasoline. For this reason, nearly all discussion of hydrogen vehicles has centered on power systems driven by fuel cells.

Fuel cells are electrochemical systems that generate electricity directly through the combination of hydrogen and oxygen in solution. Essentially electrolyzers operating in reverse, they are attractive because they have no moving parts (other than small water pumps), and under conditions where the quality of their hydrogen and oxygen feed can be perfectly controlled, they are quite efficient and reliable. These features have provided sufficient advantages to make fuel cells the technology of choice for certain specialty applications, such as the power system for NASA's Apollo capsules and the space shuttle.

Yet despite their successful use for four decades in the space program, and many billions of dollars of research and development funds expended over the years for their improvement and refinement, fuel cells have thus far found little use in broader commercial applications. The reasons for this are threefold. First, in ordinary terrestrial applications, a practical power system must last years, not just the few weeks required to support a manned space flight. Second, on Earth, the oxygen supply for the fuel cell must come from the atmosphere, which contains not only nitrogen (which decreases the fuel cell efficiency compared to a pure oxygen source), but carbon dioxide, carbon monoxide, and many other pollutants. Even in trace form, such pollutants can contaminate the catalysts used in the fuel cells and cause permanent degradation, ultimately rendering the system inoperable. Finally, and decisively, fuel cells are very expensive. For

"Natural Gas," cartoon by Stan Eales. www.CartoonStock.com. Copyright © Stan Eales. Reproduction rights obtainable from www.CartoonStock.com.

NASA, which spends hundreds of millions of dollars on every shuttle launch, it makes little difference if its 10 kilowatt fuel cell system costs $100,000, a million dollars, or ten million dollars. For a member of the public, however, such costs matter a great deal. . . .

Hydrogen Is Not a Green Fuel

Wouldn't we at least get some environmental benefit for our trillion bucks?

No, we would get no benefit at all. As discussed above, hydrogen is actually produced commercially using fossil fuel energy, much of which is lost in the process, meaning that more fossil fuels need to be burned, and thus more carbon dioxide produced, to provide a vehicle with a given amount of energy using hydrogen than if the vehicle were allowed to burn fossil fuels directly. Even if we ignore costs completely and generate hydrogen for vehicle fuel using water electrolysis, that would also *increase* pollution, since most electricity is actually generated by burning coal and natural gas. Even if the electricity in question came from nuclear, hydro, wind, or solar power, wasting it on hydrogen generation would still increase overall carbon dioxide emissions relative to the alternative of simply putting the power into the grid.

Furthermore, despite all their cost and hype, the fuel cell vehicles themselves offer no increase in efficiency relative to more conventional systems. . . .

The problem [with hydrogen] . . . is not simply economic but political, and the reality check on politicians is not always so swift or so reliable. The longer we buy into the hydrogen hoax, the longer we will avoid developing an energy policy that truly serves America's interests—economic, environmental, and geopolitical.

EVALUATING THE AUTHOR'S ARGUMENTS:

In the viewpoint you just read, the author uses facts, statistics, examples, and reasoning to make his argument that hydrogen is not likely to fuel the vehicles of the future. He does not, however, use any quotations to support his point. If you were to rewrite this article and insert quotations, what authorities might you quote from? Where would you place them, and why?

Biofuels Should Power the Vehicles of the Future

David J. Hayes, Roger Ballentine, and Jan Mazurek

"Moving America off oil by substituting homegrown biofuels for a substantial percentage of current consumption would make the country cleaner, safer, and wealthier."

In the following viewpoint David J. Hayes, Roger Ballentine, and Jan Mazurek argue that biofuels are capable of powering the vehicles of the future. They discuss how many engines are already capable of burning biofuels such as the oils of soybeans, peanuts, and cottonseed. They predict that next generation biofuels—cellulosic biofuels made from the leftover nonedible parts of crops, grasses, and trees—will make an even bigger contribution to the future fleet of vehicles. The authors support the use of biofuels because they burn cleanly, can be produced at home, and can create jobs for Americans in the process. In addition, they can use some of the fuel infrastructure that already exists (unlike hydrogen fuel cell cars, which need their own fueling stations to be built). For these and other reasons, the authors conclude that the government should support the increased use

David J. Hayes, Roger Ballentine, and Jan Mazurek, "The Promise of Biofuels," Progressive Policy Institute, March 2007. Reprinted by permission.

and development of biofuels because they are the most promising fuel of the future.

Hayes was the deputy secretary of the interior in the administration of former president Bill Clinton, for whom Ballentine was the deputy assistant to the president for environmental initiatives. Mazurek directs the Energy & Environment Project at the Progressive Policy Institute, the organization that originally published this viewpoint.

AS YOU READ, CONSIDER THE FOLLOWING QUESTIONS:
1. What kinds of biofuels do the authors say most diesel engines can already run on?
2. Why does burning biofuels not add new greenhouse gases to the atmosphere, according to the authors?
3. How many jobs do the authors say the biofuels industry helped create in 2006?

There are in fact myriad reasons to promote biofuels like ethanol, biodiesel, and the coming generation of so-called "cellulosic" variants. For starters, biofuels are *practical* alternatives to oil. Unlike, say, hydrogen fuel-cell vehicle technologies—which have only distant potential to be widely commercialized, and which would likely require a whole new service station infrastructure—expanded use of biofuels will require minimal market adaptation.

Biofuels Are Already in Use

Corn ethanol already accounts for about 3 percent of the American automotive fuel consumption. Most car engines, without any modification, can run on a blend of 90 percent gasoline and 10 percent ethanol. And carmakers have built 5 million "flex-fuel" vehicles than can run on an increasingly popular blend of just 15 percent gasoline and 85 percent ethanol, known as E85. Meanwhile, most diesel engines manufactured since 1992—including the big-rigs, tractors, and other machines that do most of the nation's heavy lifting—can run on biodiesel brewed from soybeans, peanuts, used cooking fats, animal fats, cottonseed, or canola.

Then, of course, there are the environmental benefits. Unlike gasoline made from oil, which releases carbon dioxide (CO_2) into the atmosphere when it is used in internal combustion engines, biofuels are "climate-neutral." Burning them does not add new greenhouse gases to the atmosphere, since the growth and destruction of the crops that biofuels are made from is part of the natural cycle of CO_2 absorption (during growth) and release (during destruction or decomposition).

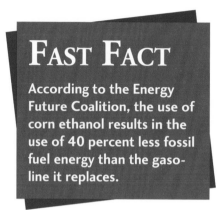

Nearly all of America's farms, rangelands, and forests, moreover, have the potential to grow plants that can be converted into biofuels. This offers the possibility of injecting new life into the U.S. agricultural sector. Even more broadly, producing fuels domestically instead of importing them from abroad will keep the profits at home, spur new investments, and create jobs—not just in the farm sector but also in processing plants and distribution systems. Industry-led studies estimate that new demand for ethanol helped create 153,725 U.S. jobs last year [in 2006]—19,000 of which were in manufacturing. Rural communities would stand to benefit the most from ethanol production because farmers own one-half of all existing ethanol refineries.

The Progressive Policy Institute (PPI) shares proponents' enthusiasm about the great promise of biofuels. But PPI believes policymakers must temper their expectations with two important caveats, which should have a direct bearing on government initiatives.

Cellulosics Are the Future of Biofuels

First, there is a natural limit to the amount of corn that U.S. farmers can grow to produce today's standard type of ethanol. At best, it is estimated that America can produce about 14 billion gallons of biofuels from corn without seriously disrupting feed and food markets. That would constitute less than 10 percent of the country's current annual motor fuel needs. The real promise of biofuels will be realized when the next generation of cellulosic biofuels can be brought to market.

Cellulosic biofuels are functionally identical from a driver's point of view to the current generation of biofuels made from corn. But they can be produced from the left-over, non-edible parts of food crops, wild grasses, and trees—which require less fertilizer, water, and energy to grow and harvest than corn. In their current state of development, cellulosic biofuels cost more than twice as much to refine, but technological breakthroughs promise to change the equation. Researchers believe they will soon be able to produce cellulosics in greater volumes, with less energy and at lower costs than corn ethanol, yielding greater net benefits in both energy and environmental terms. For now, government should certainly encourage increased production of the current generation of corn-based ethanol. But most experts agree that the real aim of such an increase in production should be to boost the supply and demand for biofuels generally, creating a ready market for cellulosic biofuels when they can be fully commercialized. . . .

The Government Should Support Biofuel Use

Owing to biofuels' great potential to help America address the steep economic, national security, and environmental costs of its oil dependence, Congress in 2005 created a new Renewable Fuels Standard (RFS). The standard—which currently applies mainly to ethanol—requires the production of 7.5 billion gallons of biofuels by 2012—and President Bush announced plans in his [2007] State of the Union address to push that target to 35 billion gallons by 2017. . . . The RFS will help to further the production of ethanol from corn and sugar in the near-term as a way to help build investor confidence in cellulosic ethanol and other advanced biofuels.

But there is more that government can do.

First and foremost, government can create the market conditions necessary for alternative fuels to compete with oil. That requires raising the price of oil to reflect its true cost to society. As it is, oil prices only reflect the direct costs of finding petroleum, pumping it out of the ground, refining it into usable fuels, and transporting it to consumers. Not included in the market price of oil are its external costs—most notably the environmental cost of *burning* it and releasing CO_2 emissions into the atmosphere. If those costs were more fully taken into account, biofuels would be much more competitive.

And there is another problem. Oil prices fluctuate wildly on global markets, to such an extent that they can undercut the appeal of alternatives. In the past year, as oil prices have at times soared past the $70 per barrel mark, biofuels have looked like a sound investment.

Invest in Biofuels

Government should also focus greater attention on research and development efforts as part of a broader effort to spur the market for clean fuels. Researchers, working with public and private backing, are currently on the cusp of technical breakthroughs that will allow efficient production of cellulosic biofuels from switchgrass, algae, and other non-edible biomass sources. The government can hasten this progress by increasing its investments in critical research projects. In the meantime, government can also goose production of the current generation of biofuels by updating the RFS and strengthening tax incentives.

Finally, in order for ethanol to be a viable gasoline substitute, it must be as cheap and easy to distribute as gasoline. America's existing system of gas pipelines cannot be used for distributing ethanol, because ethanol can corrode metal and because pipelines are not completely impervious to water. (Unlike gasoline, ethanol can absorb water, and when that happens, it becomes unsuitable as a motor fuel.) Until these problems are resolved, the nation's already-congested rail and barge networks appear to be the most likely distribution method for biofuels. At the retail end of the supply chain, an increasing number of service stations offer biofuels, but the numbers must increase if biofuels are to displace a substantial share of the nation's current gasoline consumption. Government should spur improvements on both fronts. . . .

Next-Generation Biofuels Show Great Promise

Corn-based ethanol and soy-based biodiesel already are showing great promise as a bridge to a robust market for biofuels. But the next generation of biofuels holds even greater potential to achieve the national

Biomass Resources Are Readily Available

The United States has several regions where biofuel crops can be easily grown and harvested to produce a homegrown fuel source.

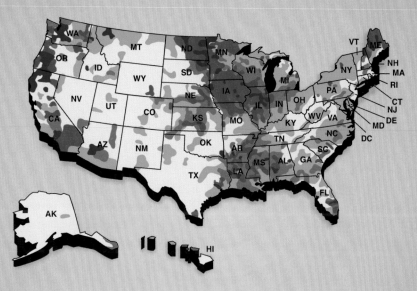

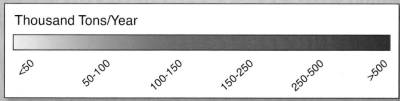

Thousand Tons/Year

<50 50-100 100-150 150-250 250-500 >500

Taken from: Milken Institute, "Financial Innovations for Energy Infrastructure: The Grid, Renewables, and Beyond," February 2010, p. 11.

objectives of energy security and reduced environmental impact. Such next-generation fuels include biodiesel from algae and waste products as well as fuels made from the non-edible parts of plants grown specifically to produce energy, not food, or from farm or forestry waste—including poplar, straw, switch grass, corn stover, and sugar cane bagasse. Moving to the next generation of biofuels is a matter of necessity. Already, the United States has more than 100 ethanol processing plants capable of producing 4.4 billion gallons of ethanol from corn annually, with another 41 under construction. Current ethanol production is taking between 13 percent and 15 percent of the current U.S. corn crop—or about 8 million acres worth—located primarily in the Midwest. At best, experts estimate that the United States has enough corn to produce between 7 billion and 14 billion gallons annually—double or triple the current production. That would represent less than 10 percent of total fuels consumed by all cars and light trucks in the United States today. If biofuels are to emerge as truly viable alternatives to gasoline, therefore, they must come from other plant sources in addition to corn. . . .

Biofuels Must Be Part of America's Energy Policy

Substituting biofuels for a substantial percentage of the gasoline and diesel fuel that Americans use in their cars and trucks will make the country safer, more prosperous, and healthier. Biofuels must therefore be central to a 21st century energy policy. . . .

Moving America off oil by substituting homegrown biofuels for a substantial percentage of current consumption would make the country cleaner, safer, and wealthier. Reducing dependence on oil and growing energy domestically can help rural economies and keep U.S. dollars at home rather than sending them to hostile regimes. Crops grown for fuel also have the potential to mitigate against global warming by keeping greenhouses gases sequestered in plants and soil rather than releasing them to the upper atmosphere where they trap heat. Despite budgetary concerns and the power of entrenched overseas oil interests, the clear and present benefits of using biofuels demand that Congress act now to spur the creation of a robust biofuels industry in America.

EVALUATING THE AUTHORS' ARGUMENTS:

The authors of this viewpoint claim that using biofuels can help avert climate change because no new greenhouse gases are released into the atmosphere when they are burned. How do you think C. Ford Runge, author of the following viewpoint, would respond to that claim? Cite examples from the texts in your answer.

Biofuels Should Not Power the Vehicles of the Future

C. Ford Runge

"Growing corn, soybeans, and other food crops to produce ethanol takes a heavy toll on the environment and is hurting the world's poor through higher food prices."

Biofuels are not the ideal fuel for the vehicles of the future argues C. Ford Runge in the following viewpoint. Runge says that biofuels are not the green fuel they have been touted to be. For one, it takes enormous amounts of land to grow biofuel crops. Trees need to be cut down to make room for these crops, which both worsens climate change and destroys natural habitat. Secondly, the fertilizers used to grow biofuel crops pollute the land and water. Finally, biofuels raise worldwide food prices, which hurts poor and hungry people around the globe. For these and other reasons, Runge concludes that biofuels are not an environmentally friendly fuel source and as such should not be supported by the US government. Runge is a professor at the University of Minnesota and is the former director of the university's Center for International Food and Agricultural Policy.

C. Ford Runge, "The Case Against Biofuels," *Yale e360*, March 11, 2010. Reprinted by permission of the author.

AS YOU READ, CONSIDER THE FOLLOWING QUESTIONS:
1. How much did the price of wheat increase between 2005 and 2008, according to Runge? What about the price of oilseeds?
2. How much more water does Runge say it takes to grow the corn that results in ethanol than is used to actually create ethanol?
3. What environmental problems are caused by the nitrogen and phosphorus fertilizers used to grow biofuels, according to Runge?

In light of the strong evidence that growing corn, soybeans, and other food crops to produce ethanol takes a heavy toll on the environment and is hurting the world's poor through higher food prices, consider this astonishing fact: This year, [2010,] more than a third of the U.S.'s record corn harvest of 335 million metric tons will be used to produce corn ethanol. What's more, within five years fully 50 percent of the U.S. corn crop is expected to wind up as biofuels.

Here's another sobering fact. Despite the record deficits facing the U.S., and notwithstanding President [Barack] Obama's embrace of some truly sustainable renewable energy policies, the president and his administration have wholeheartedly embraced corn ethanol and the tangle of government subsidies, price supports, and tariffs that underpin the entire dubious enterprise of using corn to power our cars. In early February, the president threw his weight behind new and existing initiatives to boost ethanol production from both food and nonfood sources, including supporting Congressional mandates that would triple biofuel production to 36 billion gallons by 2022.

Biofuels Are Anything but Green

Congress and the Obama administration are paying billions of dollars to producers of biofuels, with expenditures scheduled to increase steadily through 2022 and possibly 2030. The fuels are touted by these producers as a "green" solution to reliance on imported petroleum, and a boost for farmers seeking higher prices.

Yet a close look at their impact on food security and the environment—with profound effects on water, the eutrophication of our coastal zones from fertilizers, land use, and greenhouse gas emissions—suggests that the biofuel bandwagon is anything but green.

Congress and the administration need to reconsider whether they are throwing good money after bad. If the biofuel saga illustrates anything, it is that thinking ecologically will require thinking more logically, as well.

Why Biofuels Subsidies Are Wrong

Investments in biofuels have grown rapidly in the last decade, accelerating especially in developed countries and Brazil after 2003, when oil prices began to climb above $25 per barrel, reaching a peak of $120 per barrel in 2008. Between 2001 and 2008, world production of ethanol tripled from 4.9 billion gallons to 17 billion gallons, while biodiesel output rose from 264 million gallons to 2.9 billion gallons. Together, the U.S. and Brazil account for most of the world's ethanol production. Biodiesel, the other major biofuel, is produced mainly in the European Union, which makes roughly five times more than the U.S. In the EU, ethanol and biodiesel are projected to increase oilseed, wheat, and corn usage from negligible levels in 2004 to roughly

"Food for Fuel," cartoon by John Darkow, *Columbia Daily Tribune*, MO, PoliticalCartoons.com, April 16, 2008. Copyright © 2008 John Darkow and CagleCartoons.com. All rights reserved.

21, 17, and 5 million tons, respectively, in 2016, according to the Organization for Economic Cooperation and Development.

In the U.S., once a reliable supplier of exported grain and oilseeds for food, biofuel production is soaring even as food crop export demand remains strong, driving prices further upward. Government support undergirding the biofuels industry has also grown rapidly and now forms a massive federal program that may be good for farm states, but is very bad for U.S. taxpayers.

These subsidy supports are a testament to the power of the farm lobby and its sway over the U.S. Congress. In addition to longstanding crop price supports that encourage production of corn and soybeans as feedstocks, biofuels are propped up by several other forms of government largesse. The first of these are mandates, known as "renewable fuels standards": In the U.S. in 2007, energy legislation raised mandated production of biofuels to 36 billion gallons by 2022. These mandates shelter biofuels investments by guaranteeing that the demand will be there, thus encouraging oversupply.

> ## FAST FACT
>
> A 2010 study by the Institute for European Environmental Policy found that 10 million to 17 million acres of uncultivated land would need to be cleared to both grow biofuel crops and feed people in the European Union. That would release more than twice as much carbon as all of Europe's cars would produce if they continued to run on gasoline, according to the study.

Then there are direct biofuel production subsidies, which raise feedstock prices for farmers by increasing the price of corn. In the U.S., blenders are paid a 45-cent-per-gallon "blender's tax credit" for ethanol—the equivalent of more than $200 per acre to divert scarce corn from the food supply into fuel tanks. The federal government also pays a $1 credit for plant-based biodiesel and "cellulosic" ethanol.

Finally, there is a 54-cent-per-gallon tariff on imported biofuel to protect domestic production from competition, especially to prevent Brazilian sugarcane-based ethanol (which can be produced at less than half the cost of U.S. ethanol from corn) from entering U.S.

Corn that will be used in ethanol production is unloaded at a biofuel production plant. Since the introduction of biofuel, prices for wheat, rice, sugar, and oilseeds have risen dramatically worldwide.

markets. These subsidies allow ethanol producers to pay higher and higher prices for feedstocks, illustrated by the record 2008 levels of corn, soybean, and wheat prices. Projections suggest they will remain higher, assuming normal weather and yields.

Driving Up Food Prices and Increasing Hunger

The rapid increase in grain and oilseed prices due to biofuels expansion has been a shock to consumers worldwide, especially during 2008 and early 2009. From 2005 to January 2008, the global price of wheat increased 143 percent, corn by 105 percent, rice by 154 percent, sugar by 118 percent, and oilseeds by 197 percent. In 2006–2007, this rate of increase accelerated, according to the U.S. Department of Agriculture, "due to continued demand for biofuels and drought in major producing countries." The price increases have since moderated, but many believe only temporarily, given tight stocks-to-use ratios.

It is in poor countries that these price increases pose direct threats to disposable income and food security. There, the run-up in food prices has been ominous for the more than one billion of the world's poor who are chronically food-insecure. Poor farmers in countries such as Bangladesh can barely support a household on a subsistence basis, and have little if any surplus production to sell, which means they do not benefit from higher prices for corn or wheat. And poor slum-dwellers in Lagos, Calcutta, Manila, or Mexico City produce no food at all, and spend as much as 90 percent of their meager household incomes just to eat.

A Serious Toll on the Environment

But the most worrisome of recent criticisms of biofuels relate to their impacts on the natural environment. In the U.S., water shortages due to the huge volumes necessary to process grains or sugar into ethanol are not uncommon, and are amplified if these crops are irrigated. Growing corn to produce ethanol, according to a 2007 study by the U.S. National Academy of Sciences, consumes 200 times more water than the water used to process corn into ethanol.

In the cornbelt of the Upper Midwest, even more serious problems arise. Corn acreage, which expanded by over 15 percent in 2007 in response to ethanol demands, requires extensive fertilization, adding to nitrogen and phosphorus that run off into lakes and streams and eventually enter the Mississippi River watershed. This is aggravated by systems of subterranean tiles and drains—98 percent of Iowa's arable fields are tiled—that accelerate field drainage into ditches and local watersheds. As a result, loadings of nitrogen and phosphorus into the Mississippi and the Gulf of Mexico encourage algae growth, starving water bodies of oxygen needed by aquatic life and enlarging the hypoxic "dead zone" in the gulf.

Next is simply the crop acreage needed to feed the biofuels beast. A 2007 study in *Science* noted that to replace just 10 percent of the gasoline in the U.S. with ethanol and biodiesel would require 43 percent of current U.S. cropland for biofuel feedstocks. The EU would need to commit 38 percent of its cropland base. Otherwise, new lands will need to be brought into cultivation, drawn disproportionately from those more vulnerable to environmental damage, such as forests.

Biofuels Contribute to Climate Change

A pair of 2008 studies, again in *Science*, focused on the question of greenhouse gas emissions due to land-use shifts resulting from biofuels. One study said that if land is converted from rainforests, peatlands, savannas, or grasslands to produce biofuels, it causes a large net increase in greenhouse gas emissions for decades. A second study said that growing corn for ethanol in the U.S., for example, can lead to the clearing of forests and other wild lands in the developing world for food corn, which also causes a surge in greenhouse gas emissions.

A third study, by Nobel Prize–winning chemist Paul Crutzen in 2007, emphasized the impact from the heavy applications of nitrogen needed to grow expanded feedstocks of corn and rapeseed. The nitrogen necessary to grow these crops releases nitrous oxide into the atmosphere—a greenhouse gas 296 times more damaging than CO_2—and contributes more to global warming than biofuels save through fossil fuel reductions.

Thus have biofuels made the slow fade from green to brown. It is a sad irony of the biofuels experience that resource alternatives that seemed farmer-friendly and green have turned out so badly.

EVALUATING THE AUTHORS' ARGUMENTS:

C. Ford Runge and the authors of the previous viewpoint disagree on whether biofuels can rightly be called a green, or environmentally friendly, fuel. For each viewpoint, list two reasons offered for why biofuels should and should not be considered environmentally friendly. Then, state your opinion on the matter—do you think biofuels can rightly be considered an environmentally friendly fuel? Why or why not? Which piece of evidence swayed you?

Electric Cars Should Replace Gasoline-Powered Cars

"The quicker we build the necessary recharging infrastructure and start getting ourselves into electric vehicles, the better the prospects for our future."

Tom Whipple

The faster Americans transition to an electric car economy the better argues Tom Whipple in the following viewpoint. Whipple says that electric cars are superior to gasoline-powered vehicles in several ways: They use energy more efficiently, produce no polluting emissions, are easy to maintain, and perform reasonably well. Given the decline in Earth's supply of oil, Whipple says, society must be ready to transition away from gasoline-powered vehicles, and in his mind electric cars are the best alternative. For these reasons he urges Americans to invest in improving electric cars and building roads and charging infrastructure to support them. He concludes that transitioning to electric cars can protect Americans from the chaos and instability that he says will accompany the decline of oil. Whipple is a retired government analyst who writes on peak oil and related issues.

Tom Whipple, "The Peak Oil Crisis: Revisiting the Electric Car," *Energy Bulletin,* February 3, 2010. Reprinted by permission of the author.

Last week [January 2010] the US Secretary of Energy loaned Nissan motors $1.4 billion to convert an existing Nissan plant in Tennessee to build electric cars. According to the Electric Drive Transportation Association, no less than 25 models of electric cars are being readied for sale in the next few years—most by major manufacturers.

Electric Cars Are Unquestionably Superior

Although the current crop of hybrids certainly runs some of the time on electric motors, the future of electric vehicles are those that plug into the grid and get all, or at least much, of their energy from this source. There is no question that electric vehicles are intrinsically superior to the current combustion engines that have dominated personal transport for the last century. They don't use any, or not as much, petroleum-based fuels. They use energy much more efficiently. They have no emissions. Their performance is as good or better than the internal combustion car, and they are much simpler to maintain. Most places in the developed world already have robust or at least an adequate electrical distribution system for the beginning of the electric age. The last 100 feet to the car, however, will be an expensive-to-overcome problem for many.

The overwhelming advantage of the electric car is that when the time arrives that gasoline and other fossil fuels become too expensive or scarce for widespread use, the electricity probably will be there. Currently the cost of electricity per vehicle mile is very cheap in comparison with gasoline and diesel. Should the costs of electricity rise dramatically or shortages develop, many people will have the option of conserving substantial amounts of electricity at home in order to use it in an electric vehicle.

Electric Car Challenges Will Be Overcome

There are, of course, numerous downsides to the widespread adoption of the electric car and in recent weeks the press and web have been full of stories casting doubt as to their future. The most profound criticism is that the whole idea of continuing with personal vehicles in an age of declining resources—oil, coal, minerals, and personal wealth—is simply nuts. It will never happen. We are better off concentrating on public transport. Given that the world is currently running on the order of 1 billion cars and light trucks, the chances of scraping up the resources (such as lithium and car loans) to replace even a tiny fraction of such a growing fleet is unlikely.

Much of the electricity that would be used to power electric cars comes from coal and natural gas that will not be around forever. Every now and then some environmentally minded soul takes a shot at proving electric cars would make more pollution than gasoline powered ones as dirtier coals are used to generate the electricity.

Then there are a range of criticisms based on the current technical conditions pertaining to electric cars—relatively small batteries means that real-world ranges will be shorter than we are used to; most people have no convenient place to plug them in; the batteries are expensive and there will not be enough lithium to build the hundreds of millions we will need. In fact outside of North America, electric power shortages are already a problem which is likely to become worse.

Many of the smaller technical issues, however, such as range, where to recharge, and battery material have work-arounds or are likely to be overcome by the many technical advances in battery technology that are underway.

The Motorized Society Is Here to Stay

A lot of the answer as to whether we need electric vehicles lies in one's perception of what will happen to the global economy in the remaining decades of the 21st century. If one thinks the current economic difficulties will soon disappear and a new age of wealth and abundance is about to begin—then you probably won't want to buy an electric car. However, if you believe the age of petro-abundance is just about over and that oil and many other natural resources are

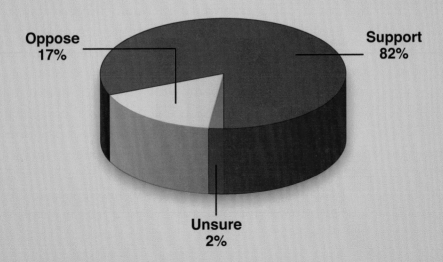

Americans Want Electric Cars

A 2009 poll found that the overwhelming majority of Americans think the government should invest in the development of electric car technology.

Question:
To address the country's energy needs, would you support or oppose action by the federal government to develop electric car technology?

Oppose
17%

Support
82%

Unsure
2%

Taken from: ABC News/*Washington Post* poll. August 13–17, 2009.

The author contends the electric car motor is more energy efficient, produces no polluting emissions, is easy to maintain, and performs reasonably well.

going to be in very short supply within the next few decades, then you should start thinking about what would be useful in the transition from the 20th century to whatever life will be like in the 22nd.

In the last 100 years the U.S. and many other parts of the world have built "motorized" societies in which life would be nearly impossible without cars and trucks. People and the entire economy move on the internal combustion engine. While returning to horses, mules, and oxen to move people and material is always possible—think of the sanitation problem and all the barns we would have to build.

If one thinks of the vast amount of infrastructure the developed world has to maintain—buildings, roads, water, sewage, trash disposal, public safety, power and communications lines—one soon gets the idea that even a relatively short range electric vehicle is going to be an awful lot better than an oxcart in preserving and rebuilding the facilities we currently rely on for food and shelter during the next 100 years or so.

Electric Cars Are Key to Our Future

There are of course ways to stretch out existing supplies of oil, perhaps for as long as a century or more, by rationing their use to only the most essential tasks needed by our civilization—farming, food transport, public safety, and utilities maintenance. This probably leaves the rest of us in the bus queue unless we find some other fuel for personal vehicles. For the immediate future, natural gas may be a substitute, but over the longer run only electricity, or possibly ammonia, made from renewable, non-polluting sources will be sustainable into future centuries.

So there is the argument. The quicker we build the necessary recharging infrastructure and start getting ourselves into electric vehicles, the better the prospects for our future.

> ## EVALUATING THE AUTHORS' ARGUMENTS:
>
> The author of this viewpoint and the author of the following viewpoint take different positions on what the point of an electric car is. Tom Whipple argues that the purpose of an electric car is to help people transition away from using fossil fuels so that they can be prepared when fossil fuels become scarce. Eric Peters argues that the point of an electric car is to help people save money on gasoline. With which of these positions do you ultimately agree? Why?

Electric Cars Are Too Expensive to Be Practical

Eric Peters

"Until someone can figure out a way to build a serviceable electric car that's priced competitively with current $11k–$15k economy cars, it's nothing more than a feel-good engineering exercise."

In the following viewpoint Eric Peters argues that electric cars are too expensive to be practical. He says the main reason people should buy electric cars is because they save on fuel costs. However, Peters points out that such savings become moot when the electric car itself costs thousands of dollars more than what a regular car costs. In his opinion, people who buy electric cars are simply spending money to save money, which he thinks does not make sense. He concludes that electric cars are only for rich people who want to make a flashy or feel-good purchase; for the average American, however, they remain too expensive to be a practical alternative vehicle.

Peters is an automotive columnist and the author of *Automotive Atrocities: The Cars You Love to Hate.*

Eric Peters, "Electric Cars and Economics 101," *American Spectator,* June 1, 2009. Reprinted by permission.

AS YOU READ, CONSIDER THE FOLLOWING QUESTIONS:
1. What does Peters say was the main problem with GM's EV-1?
2. How much does the Tesla roadster cost, according to Peters? Why does he think this makes buying the Tesla pointless?
3. Why does Peters recommend buying a Toyota Corolla or Honda Civic over any of the electric cars on today's market?

The big problem with electric cars isn't technological. It's *economic*. And one's just as defeating as the other, if the object is to come up with an electric vehicle that's more than just a cute plaything for a handful of over-rich Hollywood celebs.

Small and Limited

Consider GM's EV-1 of the 1990s.

Remember?

It was a snarky looking two-seater straight out of *Buck Rogers in the 25th Century*. Worked pretty well, too—despite what you might have read. True, its range on a single charge-up was only about a fourth of what the typical IC (internal combustion) conventional car could manage—about 70 miles or so. But that impediment was more psychological than meaningfully functional.

How many of us, after all, drive 70 miles one-way on a regular basis? Even a long commute isn't *that* long. More like 20–30 miles, for most of us. Right? And even if you push the envelope and the trip's close to the thing's max range, if it's a commute to work, you don't need to go anywhere for hours once you've parked and plugged in. Yes? So the beast would be fully recharged for the trip home.

For knocking around/commuting, the EV-1's main functional limitation was not its range; it was its two-seater configuration—and limited cargo-carrying capacity. Kind of like owning any other small two-seater, such as a BMW roadster or a Porsche.

Are Electric Cars Only for the Rich?

But the EV-1 had another thing in common with BMWs and Porsches—its *price*. Even by mid-'90s standards (when it was available to the public), the EV-1 was not a cheap date. About

In 2010 GM's electric car, the Volt, retailed for about thirty-five thousand dollars—still too expensive for the average person to afford.

$35,000—which is actually considerably more (adjusted for inflation) than the *current* base price of a 2009 model year BMW Z4 roadster ($36,700).

Now, if the whole point of owning an electric car is to save money—by saving on fuel costs—then a $35,000 electric car is as pointless, from an economy standpoint, as spending $10,000 to outfit your house with new triple pane windows but leaving them open all winter long.

Rich people don't care about gas prices. To them, $3 or $4 per gallon is like what a few extra pennies jangling in our pockets is to us. Chump change!

What's the Point of Spending to Save?

Next up, the Tesla roadster. It proves beyond any question that electric cars can also be blindingly quick and drop-dead gorgeous. It equals or beats the straight line performance (and handling) of just about any exotic sports car on the road. Keep in mind, electric motors have a huge advantage over IC engines when it comes to performance.

They produce tremendous torque *immediately*—no waiting for the RPMs to build. There is so much torque available, in fact, that a car like the Tesla can't really make full use of it—very much like a '60s-era SS 454 Chevelle on 14-inch rims.

The point, though, is not the Tesla's superb performance. It is the car's stupefying six-figure price tag—$109,000 (before taxes and tags). It's important to bear in mind that as interesting as the Tesla is, from a technology standpoint, it is basically a modified Lotus Elise roadster—which you can pick up for $46,270. The two cars perform similarly, but if you bought the Lotus, you'd still have $62,730 left over (the "change" remaining vs. the cost of the Tesla).

How much gas does $62k buy? Even at $10 per gallon, odds are you'd be in Depends before you ever had to pay for a drop of fuel again.

So, what's the point?

Like the EV-1, the Tesla is a talking point. It is sure to get you noticed. You'll be driving something "different." But you sure as heck aren't going to be saving yourself any money.

Electric Cars Are Too Expensive to Be Practical

Which brings us to the final entrant, GM's forthcoming Volt, which is a quasi-electric car in that it still has an IC engine, but it's only there to function as an onboard generator, keeping the batteries juiced up but playing no real [part] in propelling the car.

Again, very neat stuff. But once again, it's hardly economical.

GM has indicated the retail price of the Volt will be in the $30k range when it appears sometime in 2010. Maybe as much as $35k or even $40k. Which means that, like the EV-1 of the '90s, the Volt is in the same price range as

> **FAST FACT**
>
> The *Washington Examiner* reports that to make the Chevy Volt cars cost-effective, people would need to own and use them about twenty years longer than they would a regular car.

entry-luxury cars—*and therefore by definition not an economical choice, no matter how much it "saves" you on fuel.* None of that matters if the thing eats you to the bone up front, eh?

Which brings us full circle. Electric cars have proven they work. It's not the range, the batteries, the performance or the looks.

All that has been handled.

The problem is they are still too expensive to make sense for the average person. Until someone can figure out a way to build a serviceable electric car that's priced competitively with current $11k–$15k economy cars, it's nothing more than a feel-good engineering exercise.

Most people would be better advised to shop a used Corolla or Civic. They get 35 mpg without elaborate and expensive technology—and you can pick up a cherry two- or three-year-old example for $8k or so.

It may not be "green" to talk this way—but that's the truth about electric cars and Economics 101.

> ## EVALUATING THE AUTHOR'S ARGUMENTS:
>
> Peters hinges his argument about electric cars on their cost and acknowledges that this is not necessarily a "green" argument. Explain what he means by this. Then, state whether you agree—do you think electric cars should be judged on how green they are or how much they cost? Explain your reasoning.

Facts About Energy Alternatives

Editor's note: These facts can be used in reports to add credibility when making important points or claims.

Facts About Global Oil Consumption

According to the Energy Information Administration, the top ten oil-consuming nations are:

1. United States (19.5 million barrels of oil per day, or bpd)
2. China (7.8 million bpd)
3. Japan (4.8 million bpd)
4. India (3 million bpd)
5. Russia (2.9 million bpd)
6. Germany (2.6 million bpd)
7. Brazil (2.5 million bpd)
8. Saudi Arabia (2.4 million bpd)
9. Canada (2.3 million bpd)
10. South Korea (2.2 million bpd)

Together, all the nations in the world consume about 85 million barrels of oil per day.

Facts About Global Oil Production and Supply

According to the Energy Information Administration, the top ten oil producing nations are:

1. Saudi Arabia (10.8 bpd)
2. Russia (9.8 million bpd)
3. United States (8.5 million bpd)
4. Iran (4.2 million bpd)
5. China (4.0 million bpd)
6. Canada (3.4 million bpd)
7. Mexico (3.2 million bpd)
8. United Arab Emirates (3.0 million bpd)

9. Kuwait (2.7 million bpd)
10. Venezuela (2.6 million bpd)

Facts About Alternative Energy Use

According to the Energy Information Administration, as of 2010 the United States got its total energy from these sources:

Petroleum	37 percent
Natural Gas	25 percent
Coal	21 percent
Nuclear Energy	9 percent
Renewable Energy	8 percent

Renewable energy sources contributing to the 8 percent of U.S. energy listed above include:

Biomass	50 percent
Hydroelectric	35 percent
Wind	9 percent
Geothermal	5 percent
Solar	1 percent

According to the US Department of Energy, the following states and districts have pledged to make the following percentages of their total electrical energy come from renewable resources by the following years:

State	Percentage Pledged	Year
Arizona	15	2025
California	33	2030
Colorado	20	2020
Connecticut	23	2020
District of Columbia	20	2020
Delaware	20	2019
Hawaii	20	2020
Illinois	25	2025
Massachusetts	15	2020
Maryland	20	2022
Maine	40	2017

State	Percentage Pledged	Year
Michigan	10	2015
Minnesota	25	2025
Missouri	15	2021
Montana	15	2015
New Hampshire	23.8	2025
New Jersey	22.5	2021
New Mexico	20	2020
Nevada	20	2015
New York	24	2013
North Carolina	12.5	2021
North Dakota	10	2015
Oregon	25	2025
Pennsylvania	8	2020
Rhode Island	16	2019
South Dakota	10	2015
Texas	5,880 megawatts	2015
Utah	20	2025
Vermont	10	2013
Virginia	12	2022
Washington	15	2020
Wisconsin	10	2015

According to the Global Wind Energy Council, the following nations had the largest megawatt (MW) capacity for wind power:

1. United States   25,170 MW (20.8 percent of world total)
2. Germany   23,903 MW (19.8 percent)
3. Spain   16,754 MW (13.9 percent)
4. China   12,210 MW (10.1 percent)
5. India   9,645 MW (8.0 percent)
6. Italy   3,736 MW (3.1 percent)
7. France   3,404 MW (2.8 percent)
8. UK   3,241 MW (2.7 percent)
9. Denmark   3,180 MW (2.6 percent)
10. Portugal   2,862 MW (2.4 percent)

Organizations to Contact

The editors have compiled the following list of organizations concerned with the issues debated in this book. The descriptions are derived from materials provided by the organizations. All have publications or information available for interested readers. The list was compiled on the date of publication of the present volume; the information provided here may change. Be aware that many organizations take several weeks or longer to respond to inquiries, so allow as much time as possible for the receipt of requested materials.

American Solar Energy Society (ASES)
4760 Walnut St., Ste. 106
Boulder, CO 80301
(303) 443-3130
e-mail: ases@ases.org
website: www.ases.org

ASES is the nation's leading association of solar energy professionals and advocates. The group's mission is to inspire an era of energy innovation and speed the transition to a sustainable energy economy.

American Wind Energy Association (AWEA)
1501 M St. NW, Ste. 1000
Washington, DC 20005
(202) 383-2500
e-mail: windmail@awea.org
website: www.awea.org

The American Wind Energy Association represents wind power plant developers, wind turbine manufacturers, utilities, consultants, insurers, financiers, researchers, and others involved in the wind industry. The AWEA promotes the use of wind energy as a clean source of electricity for consumers around the world.

Electric Auto Association
847 Haight St.

San Francisco, CA 94117-3216
e-mail: contact@eaaev.org
website: www.eaaev.org

This group promotes the widespread adoption of plug-in electric vehicles through education and advocacy. It publishes the newsletter *Current EVents.*

International Association for Hydrogen Energy (IAHE)
5794 SW Fortieth St. #303
Miami, FL 33155
e-mail: info@iahe.org
website: www.iahe.org

The IAHE is a group of scientists and engineers professionally involved with the production and use of hydrogen. It hosts international forums to further its goal of creating an energy system based on hydrogen.

The National Renewable Energy Laboratory
1617 Cole Blvd.
Golden, CO 80401-3393
(303) 275-3000
website: www.nrel.gov

The National Renewable Energy Laboratory is the US Department of Energy's laboratory for renewable energy research, development, and deployment and a leading laboratory for studying energy efficiency. The laboratory's mission is to develop renewable energy and energy efficiency technologies and practices, advance related science and engineering, and transfer knowledge and innovations to address the nation's energy and environmental goals.

Nuclear Energy Institute
1776 Eye St. NW, Ste. 400
Washington, DC 20006-3708
(202) 739-8000
fax: (202) 785-4019
e-mail: webmasterp@nei.org
website: www.nei.org

The Nuclear Energy Institute is the policy organization of the nuclear energy industry. Its objective is to promote policies that benefit the nuclear energy business.

Renewable Energy Policy Project
1612 K St. NW, Ste. 202
Washington, DC 20006
(202) 293-2898
fax: (202) 298-5857
e-mail: info2@repp.org
website: www.repp.org

The Renewable Energy Policy Project provides information about solar, hydrogen, biomass, wind, hydrogen, and other forms of renewable energy.

Renewable Fuels Association (RFA)
425 Third St. SW, Ste. 1150
Washington, DC 20024
(202) 289-3835
e-mail: info@ethanolrfa.org
website: www.ethanolrfa.org

RFA comprises professionals who research, produce, and market renewable fuels, especially alcohol-based fuels such as ethanol.

US Environmental Protection Agency (EPA)
Ariel Rios Bldg.
1200 Pennsylvania Ave. NW
Washington, DC 20460
(202) 272-0167
website: www.epa.gov

The EPA is the federal agency in charge of protecting the environment and controlling pollution. The agency works toward these goals by enacting and enforcing regulations, identifying and fining polluters, assisting businesses and local environmental agencies, and cleaning up polluted sites.

For Further Reading

Books

Ayres, Robert U., and Edward H. Ayres. *Crossing the Energy Divide: Moving from Fossil Fuel Dependence to a Clean-Energy Future.* Upper Saddle River, NJ: Prentice-Hall, 2009. Demonstrates how existing energy systems can be reformed to maximize energy obtained from fossil fuels.

Bryce, Robert. *Power Hungry: The Myths of "Green" Energy and the Real Fuels of the Future.* New York: PublicAffairs, 2010. Argues that energy policy must be based upon four imperatives: power density, energy density, cost, and scale. Suggests that wind and solar power fail these standards.

Etherington, John. *The Wind Farm Scam.* London: Stacey International, 2009. Argues that wind power has been erroneously and excessively financed by taxpayers.

Jones, Van. *The Green Collar Economy: How One Solution Can Fix Our Two Biggest Problems.* New York: HarperOne, 2008. The author argues that environmentalism and investing in green jobs can solve the two major challenges the United States currently faces—unemployment and environmental ruin.

MacKay, David J.C. *Sustainable Energy—Without the Hot Air.* Cambridge, UK: UIT Cambridge, 2009. Proposes a sustainable energy plan for change on both a personal level and an international scale—for Europe, the United States, and the world.

McNerney, Jerry, and Martin Cheek. *Clean Energy Nation: Freeing America from the Tyranny of Fossil Fuels.* New York: AMACOM, 2011. Explores options for replacing fossil fuels and examines who should bear responsibility for developing a sustainable energy economy.

Richter, Burton. *Beyond Smoke and Mirrors: Climate Change and Energy in the 21st Century.* New York: Cambridge University Press, 2010. Presents various options for moving away from heavy reliance on fossil fuels to a more sustainable energy system.

Smil, Vaclav. *Energy Myths and Realities: Bringing Science to the Energy Policy Debate.* Washington, DC: American Enterprise Institute, 2010. Debunks the most common fallacies to make way for a constructive, scientific approach to the global energy challenge.

Periodicals and Internet Sources

American Wind Energy Association. "Wind Power: Myths vs. Facts." www.awea.org/pubs/factsheets/050629-Myths vs Facts Fact-Sheet. pdf.

Boudreaux, Donald J. "For Oil, Tap Ingenuity," *Pittsburgh Tribune,* February 24, 2010. www.pittsburghlive.com/x/pittsburghtrib/ opinion/columnists/boudreaux/s_668583.html.

Breining, Greg. "From the Sewage Plant, the Promise of Biofuel," *Yale Environment360,* July 1, 2009. http://e360.yale.edu/content/ feature.msp?id=2167.

Bryce, Robert. "The Brewing Tempest over Wind Power," *Wall Street Journal,* March 2, 2010. www.manhattan-institute.org/html/miar ticle.htm?id=6066.

———. "The Real Problem with Renewables," *Forbes,* May 11, 2010. www.manhattan-institute.org/html/miarticle.htm?id=6223.

———. "Windmills Are Killing Our Birds," *Wall Street Journal,* September 7, 2009. http://online.wsj.com/article/SB1000142405 2970203706604574376543308399048.html.

Carrion, Fabiola. "Clean Energy Options: In the Wake of the Oil Spill, Energy Alternatives That Will Create Jobs," Progressive States Network, July 19, 2010. http://progressivestates.org/node/25318.

Chameides, Bill. "Biofuels Part I: Corn Ethanol Isn't the Solution," The Green Grok, Duke University, May 21, 2008. http://nicholas. duke.edu/thegreengrok/greenoptions4.30.

Connor, Steve. "Warning: Oil Supplies Are Running Out Fast," *Independent* (London), August 3, 2009. www.independent.co.uk/ news/science/warning-oil-supplies-are-running-out-fast-1766585 .html.

Energy Independence Institute. "America's Solar Energy Potential." http://americanenergyindependence.com/solarenergy.aspx.

Garber, Kent. "The Green Energy Economy: What It Will Take to Get There," *U.S. News & World Report,* March 20, 2009. http:// politics.usnews.com/news/energy/articles/2009/03/20/the-green- energy-economy-what-it-will-take-to-get-there print.html.

Goodall, Chris. "The 10 Big Energy Myths," *Guardian* (Manchester, UK), November 27, 2008. www.guardian.co.uk/environment/2008/nov/27/renewableenergy-energy.

Gore, Al. "The Crisis Comes Ashore," *New Republic*, May 9, 2010. www.michaelmoore.com/words/must-read/crisis-comes-ashore.

Grunwald, Michael. "Seven Myths About Alternative Energy," *Foreign Policy*, September/October, 2009. www.foreignpolicy.com/articles/2009/08/12/seven myths about alternative_energy.

Hanlon, Michael. "Time to Unplug the Electric Car," *Daily Mail* (London), June 7, 2010. http://hanlonblog.dailymail.co.uk/2010/06/time-to-unplug-the-electric-car.html.

Hill, Patrice. "Green Energy Stimulus Growing Few Jobs," *Washington Times*, November 23, 2009. www.washingtontimes.com/news/2009/nov/23/green-stimulus-growing-few-jobs/print.

———. "'Green' Jobs No Longer Golden in Stimulus," *Washington Times*, September 8, 2010. www.washingtontimes.com/news/2010/sep/9/green-jobs-no-longer-golden-in-stimulus/?page=1.

Leggett, Jeremy. "Society Ignores the Oil Crunch at Its Peril," *Guardian* (Manchester, UK), February 10, 2010. www.guardian.co.uk/environment/cif-green/2010/feb/10/oil-crunch-peril.

Lomborg, Bjørn. "Cheap Green Energy: The Only Way to Fight Global Warming," *Daily Star* (Beirut, Lebanon), July 23, 2010. www.dailystar.com.lb/article.asp?edition-id=10&categid=5&article id=117363#axzz0uWr7dPwd.

Milligan, Michael, et al. "Wind Power Myths Debunked," *IEEE Power & Energy Magazine*, November/December 2009. www.ieee-pes.org/images/pdf/open-access-milligan.pdf.

Morris, David. "Give Ethanol a Chance: The Case for Corn-Based Fuel," AlterNet, July 13, 2007. www.newrules.org/energy/article/give-ethanol-chance-case-cornbased-fuel.

Nature Conservancy. "Climate Change and Energy: The True Cost of Biofuels," 2009. www.nature.org/initiatives/climatechange/features/art23819.html.

Pagano, Margareta. "Are Wind Farms a Health Risk?" *Independent* (London), August 2, 2009. www.independent.co.uk/environment/green-living/are-wind-farms-a-health-risk-us-scientist-identifies-wind-turbine-syndrome-1766254.html.

Pickens, T. Boone, and Ted Turner. "Natural Gas and Renewables Are the Key to a Cleaner, More Secure Energy Supply," *Wall Street Journal*, August 16, 2009. http://online.wsj.com/arti cle/SB10001424052970203863204574348432504983734 .html?mod=djemEditorialPage.

Rose, David. "Has the Spark Gone Out of Electric Cars?," *Daily Mail* (London), August 22, 2010. www.dailymail.co.uk/home/moslive/ article-1304123/Has-spark-gone-electric-cars.html.

Runge, C. Ford, and Benjamin Senauer. "How Biofuels Could Starve the Poor," *Foreign Affairs*, May/June 2007. www.foreignaffairs. com/articles/62609/c-ford-runge-and-benjamin-senauer/how- biofuels-could-starve-the-poor.

Sanders, Bernie. "It's Time for a Solar Revolution," February 11, 2010. http://sanders.senate.gov/newsroom/news/?id=867f0eb5- e6d0-42ef-aa25-13bc3b7ce754.

Schmidt, Uwe T. "Oil Spill Underscores the Need for Clean Energy Alternatives," *Palm Springs (CA) Desert Sun*, July 2, 2010. http:// pqasb.pqarchiver.com/mydesert/access/2074760051.html?FMT= ABS&date=Jul+02%2C+2010.

Sherraden, Samuel. "Green Jobs: Hope or Hype?," CNN.com, July 28, 2009. http://articles.cnn.com/2009-07-28/politics/sher raden.green.jobs_1_green-jobs-green-sector-energy-efficiency?_ s=PM:POLITICS.

Steinmetz, Tara. "Nuclear Energy Is Not a Viable Alternative to Oil Dependency," *Charlotte (NC) Observer*, July 21, 2010. www.char lotteobserver.com/2010/07/21/1573786/nuclear:energy-is-not-a- viable.html.

Stelzer, Irwin. "It's a Myth That the World's Oil Is Running Out," *Times* (London), April 27, 2008. http://business.timesonline. co.uk/tol/business/columnists/article3823656.ece.

Tollefson, Jeff. "Hydrogen Vehicles: Fuel of the Future?," *Nature*, April 29, 2010. www.nature.com/news/2010/100429/full/4641262a .html?s=news_rss.

Tsur, Doron. "You Want Green Energy? Pay with High Unemployment," *Ha'aretz* (Tel Aviv, Israel), June 24, 2010. www .haaretz.com/print-edition/business/you-want-green-energy-pay- with-high-unemployment-1.297976.

Walsch, Bryan. "Tallying Biofuels' Real Environmental Cost," *Time*, October 23, 2009. www.time.com/time/health/article/0,8599,1931780,00.html.

Will, George F. "Tilting at Green Windmills," *Washington Post*, June 25, 2009. www.washingtonpost.com/wp-dyn/content/article/2009/06/24/AR2009062403012.html.

Wilson, Andrew B. "One Job Forward, Two Jobs Back," *Weekly Standard*, July 19, 2010. www.weeklystandard.com/articles/one-job-forward-two-jobs-back?page=1.

Zubrin, Robert. "In Defense of Biofuels," *New Atlantis*, Spring 2008. www.thenewatlantis.com/publications/in-defense-of-biofuels.

Websites

Alternative Energy Resources Organization (AERO) (www.aeromt.org). This is the website of a grassroots nonprofit organization dedicated to solutions that promote resource conservation and local economic vitality. AERO nurtures individual and community self-reliance through programs that support sustainable agriculture, renewable energy, and environmental quality. The site features information about alternative energy resources and agriculture.

Energy Bulletin (www.energybulletin.net). This site, published by the Post Carbon Institute, serves as a clearinghouse for information about the global energy supply. Abundant information on energy alternatives can be found here.

Intergovernmental Renewable Energy Organization (IREO) (www. ireoigo.org). This group works to promote the use of affordable, clean sources of renewable energy worldwide.

US DOE Energy Efficiency and Renewable Energy (EERE) Home Page (www.eere.energy.gov). This government site offers information about the US Department of Energy's Office of Energy Efficiency and Renewable Energy. EERE invests in clean energy technologies that strengthen the economy, protect the environment, and reduce dependence on foreign oil.

Index

are too expensive to be practical, 130–134
competition for lithium and, 44, 47–48
motor of, *128*
should replace gasoline-powered cars, 124–129
support for, *127*
Electric Drive Transportation Association, 125
Electricity
amount generated by wind turbines, 64
amount produced by nuclear power in US, 77–78, 84
energy equivalent of, in oil, 60–61
percent provided by hydro power in US, 79
potential of photovoltaic cells in production of, 51
Emissions
from hydrogen fuel cell vehicles, 101
See also Carbon dioxide; Greenhouse gases
Energy, 71
Energy consumption, *59*
Energy Future Coalition, 112
Energy Information Administration, US, 59, 60, 84
Environmental engineers, 7–8
Ethanol
cellulosic, 112–113
corn, 111
distribution of, 114
major producers of, 119

percent of US corn crop used for, 116–117, 118
Europe
nuclear power in, 78
offshore wind energy in, 67
potential of solar power in, 71–72
EV-1 (electric car), 131–132
Everley, Steve, 20

F
Farrell, John, 50
Fossil fuels
dependence on, 38–42
See also specific types
Freese, Charlie, 102

G
Gallucci, Robert, 45
Gartenstein-Ross, Daveed, 38
Geological Survey, US (USGS), 22
Ghawar oil field (Saudi Arabia), 22
Gingrich, Newt, 20
Global Wind Energy Council, 27
Gnacadja, Luc, 28
Gore, Al, 22
Government, US
amount spent subsidizing nuclear power, 92
should not subsidize biofuel production, 119–121
should support biofuel use, 113–114
support for alternative energy research by, *29*

transportation/storage of, 90,
91, *94*

O
Obama, Barack/Obama
administration, 45
cuts funding for hydrogen
auto research, 99–100
stimulus package of, 8
support for renewable energy,
58–59, 61–62
support of corn ethanol, 118
OECD (Organization for
Economic Cooperation and
Development), 83, 120
Office of Technology
Assessment, US, 94
Offshore wind turbines, 67
Ogden, Joan, 98, 102
Oil/oil reserves
are not running out, 20–24
are running out, 12–19
external costs are not included
in price of, 113–114
per-barrel energy equivalent
in electricity, 60
Organization for Economic
Cooperation and
Development (OECD), 83,
120
Ostrander, Madeline, 25
Oxfam International, 33, 37
Oxford Institute of Energy
Studies, 14–15

P
Partrick, Neil, 40
Peak oil theory, 17, 20, 21–22

Peters, Eric, 130
Photovoltaic solar panels, *52,* 60
Pibel, Doug, 25
Popular Mechanics (magazine),
126
Progressive Policy Institute
(PPI), 112
Progressive States Network, 8
Protection technicians, 8
Public Agenda, 30

R
Rainforests, 34
Renewable energy
can meet energy demands of
US, 50–56
cannot meet energy demands
of US, 57–62
states with goals/standards
for, *55*
subsidies for, 9
Renewable Fuels Standard
(RFS), 113, 114
Rothkopf, David, 43
Runge, C. Ford, 117

S
Salazar, Ken, 21
Saudi Arabia, 40–41
Sawt al-Jihad (online magazine),
39
Schulz, Max, 9
Science (journal), 122
Scientific American (magazine),
15, 73
Sellafield nuclear reprocessing
plant (UK), *86*
Sherraden, Samuel, 8, 9

Picture Credits

AP Images/Douglas Engle, 21

AP Images/Kevin Rivoli, 35

AP Images/Michael Prost, 94

AP Images/Nabil al-Jurani, 41

AP Images/Paul Sancya, 132

AP Images/Rick Bowmer, 61

AP Images/Solar Systems, HO, 72

Martin Bond/Photo Researchers, Inc., 52, 100

Peter Bowater/Photo Researchers, Inc., 11, 15

Gaston Brito/Reuters/Landov, 46

Claire Deprez/Reporters/Photo Researchers, Inc., 49

Gale, Cengage Learning, 13, 16, 23, 29, 36, 55, 59, 66, 74, 85, 99,
     115, 127

Jerry Mason/Photo Researchers, Inc., 96

Phil Noble/PA Photos/Landov, 86

Philippe Psaila/Photo Researchers, Inc., 128

Jason Reed/Reuters/Landov, 121

Bob Strong/Reuters/Landov, 68